AF588392

ESAFORM Bookseries on Material Forming

Series editor

Elías Cueto, Zaragoza, Spain

For further volumes:
http://www.springer.com/series/11923

Francisco Chinesta · Elías Cueto

PGD-Based Modeling of Materials, Structures and Processes

Springer

Francisco Chinesta
UMR CNRS
Ecole Centrale de Nantes
Nantes
France

Elías Cueto
Aragon Institute of Engineering Research
Universidad de Zaragoza
Zaragoza
Spain

ISSN 2198-9907 ISSN 2198-9915 (electronic)
ISBN 978-3-319-06181-8 ISBN 978-3-319-06182-5 (eBook)
DOI 10.1007/978-3-319-06182-5
Springer Cham Heidelberg New York Dordrecht London

Library of Congress Control Number: 2014937278

Printed on acid-free paper

Springer is part of Springer Science+Business Media (www.springer.com)

To Ofelia and Susana

Preface

In this monograph we address a new paradigm in the field of simulation-based engineering sciences (SBES) to face the challenges posed by current ICT technologies.

In materials, processes, and structural design many scenarios must be considered and carefully analyzed, and this task is expensive from a computational point of view. Moreover, the design parametric space is too large to consider its exhaustive exploration. In general, only coarse samplings of the parametric spaces are performed, complemented with the use of design security coefficients for including the unknown information and also the inevitable uncertainty. Thus, in practice, designers consider very well-tested materials, manufacturing processes, and loading scenarios in order to guarantee a smooth design process (fine and expensive simulations are restricted to the analysis of few critical designs) and in consequence the resulting designs remain suboptimal.

Nowadays, industry is using materials and manufacturing processes that are, for most of them, 30-years old. New designs, and more precisely, a new design framework is urgently needed, allowing us to explore regions of the design space never until now explored, making possible breakthroughs in both the design (in its largest sense) and the product technologies. Moreover, this new framework should allow us to address uncertainty quantification and its propagation throughout the whole design chain (including processes and experimental tests). It should also allow addressing efficiently high-fidelity models developed in the last quarter of the century but until now inefficiently considered because of their complexity for current decision-making strategies.

Real-time analysis of complex systems is compulsory for making possible real-time decision-making that needs the evaluation of many possible scenarios under the real-time constraint. Decision making is at the heart of material, processes, and structural optimization and also of the incipient simulation-based control. Moreover, for democratizing accessibility to efficient design technologies, decision-making tools should run in light computing devices.

These apparently contradictory requirements, the real-time evaluation of system responses based on high-fidelity models and involved in decision-making tools and the suitability of running these applications and tools in light computational devices, could be possible if we generate offline a sort of a computational vademecum containing the solution of the model under consideration for all possible design scenarios and then use it online for decision-making purposes.

We developed in recent years a novel technique, called Proper Generalized Decomposition (PGD). It is based on the assumption of a separated form of the unknown field and it has demonstrated its capabilities in dealing with high-dimensional problems overcoming the strong limitations of classical approaches. Many challenging problems can be efficiently cast into a multidimensional framework. For instance, parameters in a model (loads, initial conditions, boundary conditions, material parameters, geometrical parameters, etc.) can be set as additional extra-coordinates of the model. In a PGD framework, the resulting model is solved once for life, in order to obtain a general solution that includes all the solutions for every possible value of the parameters, that is, a sort of computational vademecum. Under this rationale, optimization of complex problems, uncertainty quantification, simulation-based control, and real-time simulation are now at hand in highly complex scenarios and on deployed platforms.

Consider a material, process, or structural model, or one coupling these three items. This model has an associated output (displacement field, stress field, temperature field, electric or magnetic fields, etc.) of engineering interest. However, this output depends on several parameters that were fixed before solving it (e.g., position and magnitude of the applied loads, boundary conditions, initial conditions, material parameters—young modulus, thermal conductivity, etc.—or geometrical parameters associated with the system under consideration). Thus, for each possible scenario or tentative design we must solve the model, a very costly task in practice that makes unaffordable a design (in its largest sense) based on a high-fidelity modeling on the whole parametric space domain.

For a given family of problems, imagine now, that we consider all the model parameters as coordinates (as the space and time are in standard models). Now, by solving offline and only once the resulting multidimensional model, we have access to the output (the fields of engineering interest) for any value of the design parameters (e.g. position and magnitude of the loads, boundary or initial conditions, material or geometry). By considering this multiparametric solution, a sort of modern "vademecum," design (in its largest sense) becomes extremely efficient because designers can explore in real-time and on light computing platforms, the whole parametric design space related to the high-fidelity model solution. We can visualize this multiparametric solution as a sort of virtual chart, a design facility that makes possible fast and extremely accurate optimization, inverse analysis, control, and uncertainty propagation that only requires particularizing the multiparametric solution as many times as required.

PGD also allows alleviating the solution of complex 3D models defined in degenerated domains (plate and shell-like domains), the solution of transient models in a non-incremental way, and the solution of models defined in a multi-dimensional domain, those in general making use of a number of conformation coordinates as encountered in the fine modeling of the mechanics of materials, most of them revisited in this monograph.

Nantes, December 2013 Francisco Chinesta
Zaragoza Elías Cueto

Acknowledgments

We are deeply indebted to those who did most part of the work included here. Of course, they are responsible for the good results presented here, and not for the mistakes. We mean all the members of our labs at Nantes and Zaragoza, Ph.D. students, postdocs, and associate professors who have worked hard to make PGD come true as a valuable tool in many fields of applied science and engineering and, particularly, in the world of forming processes.

Our gratitude goes to (in alphabetical order) the members of our research groups: José Vicente Aguado, Icíar Alfaro, Brice Bognet, Felipe Bordeu, Chady Ghnatios, David González, Adrien Leygue, Siamak Niroomandi, Fabien Poulhaon, Fanou Retat-Masson, and so on among many others who also contributed to the works presented here.

A special acknowledgement is deserved by professors Amine Ammar, Antonio Huerta, and Roland Keunings, who inspired and participated in most of the applications included here. It has been a pleasure to work with all of them, and we are confident of a fruitful future of joint work.

Contents

Chapter 1
Introduction

Imagination is more important than knowledge.
—Albert Einstein

1.1 Recurrent Issues in Numerical Simulation

Many problems in science and engineering remain intractable nowadays because of their numerical complexity, or the restrictions imposed by different requirements (real-time on deployed platforms, for instance) make them unaffordable for today's technologies.

In this monograph we are focussing on five scenarios:

- The first one concerns models that are defined in high dimensional spaces.
  These models are usually encountered in quantum chemistry describing the structure and mechanics of materials [1]; the kinetic theory description of complex fluids [2] or chemical modeling in too dilute systems where the concept of concentration cannot be used, that results in the so-called chemical master equation [3]. Models defined in high dimensional spaces suffer the so-called curse of dimensionality. If one proceeds to the solution of a model defined in a space of dimension D by using a standard mesh based discretization technique, where M nodes are used for discretizing each space coordinate, the resulting number of nodes reaches the astronomical value of M^D. With $M \approx 10^3$ (a very coarse description in practice) and $D \approx 30$ (a very simple model) the numerical complexity results to be on the order of 10^{90}, which corresponds to the presumed number of elementary particles in the universe.
- The second one concerns transient coupled models with different characteristic times.
  When problems combine different physics with different characteristic times two difficulties arise. The first concerns the transient simulation itself when the time

F. Chinesta and E. Cueto, *PGD-Based Modeling of Materials, Structures and Processes*,
ESAFORM Bookseries on Material Forming, DOI: 10.1007/978-3-319-06182-5_1,

step becomes too small with respect to the time interval of interest. This is the situation found when analyzing material aging. The second difficulty concerns the space-time coupling of such multi scale models.

- The third concerns models defined in degenerated domains.
Plates or shells constitute degenerated geometries because one of its characteristic dimensions (thickness) is much lower that the other characteristic dimensions. In the case of beams two characteristic dimensions are much lower than the one related to the beam length. When a model is defined in such kind of geometries its 3D numerical solution becomes delicate. In the numerical framework the solution will be only obtained in a discrete number of points, usually called nodes, distributed in the domain. From the solution at those points, it can be interpolated at any other point in the domain. In general regular nodal distributions are preferred because they offer the best accuracy. In the case of degenerated plate or shell domains one could expect that if the solution evolves significantly in the thickness direction, a large enough number of nodes must be distributed along the thickness direction to ensure the accurate representation of the field evolution in that direction. In that case, a regular nodal distribution in the whole domain will imply the use of an extremely large number of nodes with the consequent impact on the numerical solution efficiency.
- The fourth concerns applications needing real time responses.
Online control is usually performed by considering a system as a black box whose behavior is modeled by a transfer function relating certain inputs to certain outputs. This modeling that may seem poor has one main advantage, the possibility of proceeding rapidly due to its simplicity. The establishment of such a goal-oriented transfer function is the trickiest point. Actual physical models result, in general, in complex mathematical objects, nonlinear and strongly coupled partial differential equations. Such mathematical objects are representing physical reality up to a certain degree of accuracy. However, the available numerical tools capable of solving these complex models require the use of powerful computers that can require hours, days and weeks to solve them. Known as numerical simulation, its output solution is very rich but it seems inapplicable for control purposes that require fast responses, often in real-time.
Simulation-based Engineering Sciences (SBES) relied on the use of static data inputs to perform the simulations. These data could be parameters of the model(s) or boundary conditions. The word static is intended to mean here that these data could not be modified during the simulation. A new paradigm in the field of Applied Sciences and Engineering has emerged in the last decade. Dynamic Data-Driven Application Systems (DDDAS) constitute nowadays one of the most challenging applications of Simulation-Based Engineering Sciences. By DDDAS we mean a set of techniques that allow the linkage of simulation tools with measurement devices for real-time control of simulations. DDDAS entails the ability to dynamically incorporate additional data into an executing application, and in reverse, the ability of an application to dynamically steer the measurement process [4–6]. In this context, real time simulators are needed in many applications.

Augmentedreality is another area in which efficient (fast and accurate) simulation is urgently needed. The idea is supplying in real time appropriate information to the reality perceived by the user. Augmented reality could be an excellent tool in many branches of science and engineering. In this context, light computing platforms are appealing alternatives to heavy computing platforms that in general are expensive and whose use requires technical knowledge.

- The last one concerns models that must be solved many times.
  Many problems in parametric modeling, inverse identification, and process or shape optimization, usually require, when approached with standard techniques, the direct computation of a very large number of solutions of the concerned model for particular values of the problem parameters. When the number of parameters increases such a procedure becomes inapplicable.
  Moreover in science and engineering, in its widest sense, it now seems obvious that there are many causes of variability. The introduction of such variability, randomness and uncertainty is a priority for the next decade. Although it was a priority in the preceding decade, the practical progress attained seems fairly weak.

While the previous list is by no means exhaustive, it includes a set of problems with no apparent relationship between them that can however be treated in a unified manner as will be shown in what follows. Their common ingredient is our lack of capabilities (or knowledge) to solve them numerically in a direct, traditional way. In order to obtain a solution, some kind of model order reduction is thus compulsory.

We recently have developed a new generation of simulation strategies, which consist essentially in a PGD-based Reduced Order Modelling (ROM) strategy. We first describe the construction of such an approximation and then illustrate how this approximation could constitute a new paradigm in computational sciences allowing to circumvent the above computational issues.

1.2 Model Reduction: Information Versus Relevant Information

Model Order Reduction techniques have been extensively used for solving transient models defined in space domains involving a large number of degrees of freedom, small time steps and a large time interval.

Imagine a model whose solution $u(\mathbf{x}, t)$ is defined in $\mathbf{x} \in \Omega$ with $t \in \mathscr{I} = (0, T]$. By using an appropriate discretization technique the solution is calculated at M points $\mathbf{x}_i, i = 1, \cdots, M$ and P time instants $t_m, m = 1, \cdots, P$, being the number of nodes M and time instants P large enough for ensuring convergence and stability of the discrete solution.

Standard discretization techniques (finite elements, finite volumes, finite differences, meshless techniques,) require in general the solution of a linear system of size M at each time instant. Thus the computational complexity scales with $P \cdot M$. When M becomes large because of the solution localization or the domain complexity, or/and P because of stability constraints or because of the size of the time interval with respect to the characteristic time of the physics involved, the resulting complexity $P \cdot M$ could become unaffordable.

Model reduction techniques based on the use of Proper Orthogonal Decomposition allowed circumventing the just cited computational difficulty when the problem solution lives in a subspace of dimension much smaller than M.

1.2.1 Extracting Relevant Information: Proper Orthogonal Decomposition

We assume that the field of interest $u(\mathbf{x}, t)$ is known at the nodes $\mathbf{x}_i$ of a spatial mesh for discrete times $t_m = m \cdot \Delta t$, with $i \in [1, \ldots, M]$ and $m \in [0, \ldots P]$. We use the notation $u(\mathbf{x}_i, t_m) \equiv u^m(\mathbf{x}_i) \equiv u_i^m$ and define $\mathbf{u}^m$ as the vector of nodal values u_i^m at time t_m. The main objective of the POD is to obtain the most typical or characteristic structure $\phi(\mathbf{x})$ among these $u^m(\mathbf{x})$, $\forall m$ [7]. For this purpose, we maximize the scalar quantity

$$\alpha = \frac{\sum_{m=1}^{P} \left[\sum_{i=1}^{M} \phi(\mathbf{x}_i) u^m(\mathbf{x}_i) \right]^2}{\sum_{i=1}^{M} (\phi(\mathbf{x}_i))^2}, \tag{1.1}$$

which amounts to solving the following eigenvalue problem:

$$\mathbf{C}\phi = \alpha\phi. \tag{1.2}$$

Here, the vector ϕ has i-component $\phi(\mathbf{x}_i)$, and $\mathbf{c}$ is the two-point correlation matrix

$$\mathbf{C}_{ij} = \sum_{m=1}^{P} u^m(\mathbf{x}_i) u^m(\mathbf{x}_j) \tag{1.3}$$

whose compact form is written

$$\mathbf{C} = \sum_{m=1}^{P} \mathbf{u}^m \cdot (\mathbf{u}^m)^T, \tag{1.4}$$

which is symmetric and positive definite. With the matrix $\mathbf{Q}$ defined as

$$\mathbf{Q} = \begin{pmatrix} u_1^1 & u_1^2 & \cdots & u_1^P \\ u_2^1 & u_2^2 & \cdots & u_2^P \\ \vdots & \vdots & \ddots & \vdots \\ u_M^1 & u_M^2 & \cdots & u_M^P \end{pmatrix}, \tag{1.5}$$

we have

$$\mathbf{C} = \mathbf{Q} \cdot \mathbf{Q}^T. \tag{1.6}$$

In order to obtain a reduced model, we first solve the eigenvalue problem Eq. (1.2) and select the N eigenvectors ϕ_i associated with the eigenvalues belonging to the interval defined by the highest eigenvalue α_1 and α_1 divided by a large enough number (e.g. 10^8). In practice, for many models, N is found to be much lower than M. These N eigenfunctions ϕ_i are then used to approximate the solution $u^m(\mathbf{x})$, $\forall m$. To this end, let us define the matrix $\mathbf{B} = [\phi_1 \cdots \phi_N]$, i.e.,

$$\mathbf{B} = \begin{pmatrix} \phi_1(\mathbf{x}_1) & \phi_2(\mathbf{x}_1) & \cdots & \phi_N(\mathbf{x}_1) \\ \phi_1(\mathbf{x}_2) & \phi_2(\mathbf{x}_2) & \cdots & \phi_N(\mathbf{x}_2) \\ \vdots & \vdots & \ddots & \vdots \\ \phi_1(\mathbf{x}_M) & \phi_2(\mathbf{x}_M) & \cdots & \phi_N(\mathbf{x}_M) \end{pmatrix}. \tag{1.7}$$

Now, let us assume for illustrative purposes that an explicit time-stepping scheme is used to compute the discrete solution $\mathbf{u}^{m+1}$ at time t^{m+1}. One must thus solve a linear algebraic system like

$$\mathbf{G}^m \ \mathbf{u}^{m+1} = \mathbf{H}^m. \tag{1.8}$$

A reduced-order model is then obtained by approximating $\mathbf{u}^{m+1}$ in the subspace defined by the N eigenvectors ϕ_i, i.e.,

$$\mathbf{u}^{m+1} \approx \sum_{i=1}^{N} \phi_i \ T_i^{m+1} = \mathbf{B} \ \mathbf{T}^{m+1}. \tag{1.9}$$

Equation (1.8) then reads

$$\mathbf{G}^m \ \mathbf{B} \ \mathbf{T}^{m+1} = \mathbf{H}^m, \tag{1.10}$$

or equivalently

$$\mathbf{B}^T \mathbf{G}^m \ \mathbf{B} \ \mathbf{T}^{m+1} = \mathbf{B}^T \mathbf{H}^m. \tag{1.11}$$

The coefficients T_i^{m+1} defining the solution of the reduced-order model are thus obtained by solving an algebraic system of size N instead of M. When $N \ll M$, as is the case in numerous applications, the solution of Eq. (1.11) is thus preferred because of its much reduced size.

Remark 1.2.1 The reduced-order model Eq. (1.11) is built a posteriori by means of the already-computed discrete field evolution. Thus, one could wonder about the interest of the whole exercise. In fact, two beneficial approaches are widely considered (see e.g., [7–14]). The first approach consists in solving the large original model over a short time interval, thus allowing for the extraction of the characteristic structure that defines the reduced model. The latter is then solved over larger time intervals, with the associated computing time savings. The other approach consists in solving the original model over the entire time interval, and then using the corresponding reduced model to efficiently solve similar problems with, for example, slight variations in material parameters or boundary conditions.

1.2.2 Towards an "A Priori" Space-Time Separated Representation

The above example illustrates the significant value of model reduction. Of course, one would ideally want to be able to build a reduced-order approximation *a priori*, i.e. without relying on the knowledge of the (approximate) solution of the complete problem. One would then want to be able to assess the accuracy of the reduced-order solution and, if necessary, to enrich the reduced approximation basis in order to improve accuracy (see e.g., our earlier studies [15] and [7]). The Proper Generalized Decomposition (PGD), which we describe in general terms in the next section, is an efficient answer to these questions.

The above POD results also tell us that an accurate approximate solution can often be written as a separated representation involving but few terms. Indeed, when the field evolves smoothly, the magnitude of the (ordered) eigenvalues α_i decreases very fast with increasing index i, and the evolution of the field can be approximated from a reduced number of modes. Thus, if we define a cutoff value ε (e.g., $\varepsilon = 10^{-8} \cdot \alpha_1$, α_1 being the highest eigenvalue), only a small number N of modes are retained ($N \ll M$) such that $\alpha_i \geq \varepsilon$, for $i \leqslant N$, and $\alpha_i < \varepsilon$, for $i > N$. Thus, one can write:

$$u(\mathbf{x}, t) \approx \sum_{i=1}^{N} \phi_i(\mathbf{x}) \cdot T_i(t) \equiv \sum_{i=1}^{N} X_i(\mathbf{x}) \cdot T_i(t). \tag{1.12}$$

For the sake of clarity, the space modes $\phi_i(\mathbf{x})$ will be denoted in the sequel as $X_i(\mathbf{x})$. Equation (1.12) represents a natural *separated representation*, also known as finite sum decomposition. The solution that depends on space and time can be approximated as a sum of a *small number* of functional products, with one of the functions depending on the space coordinates and the other one on time. Use of separated representations like (1.12) is at the heart of the PGD.

To our knowledge, the unique precedent to the PGD algorithm for building a separated space-time representation is the so-called radial approximation introduced by Ladeveze [16–18] in the context of Computational Solid Mechanics.

In terms of performance, the verdict is simply impressive. Consider a typical transient problem defined in 3D physical space. Use of a standard incremental strategy with P time steps (P could be of order of millions in industrial applications) requires the solution of P three-dimensional problems. By contrast, using the space-time separated representation (1.12), we must solve $N \cdot m$ three-dimensional problems for computing the space functions $X_i(\mathbf{x})$, and $N \cdot m$ one-dimensional problems for computing the time functions $T_i(t)$. Here, m is the number of nonlinear iterations needed for computing each term of the finite sum. For many problems of practical interest, we find that $N \cdot m$ is of order 100. The computing time savings afforded by the separated representation can thus reach many orders of magnitude.

1.3 The Proper Generalized Decomposition at a Glance

Consider a problem defined in a space of dimension D for the unknown field $u(x_1, \ldots, x_D)$. Here, the coordinates x_i denote any usual coordinate (scalar or vectorial) related to physical space, time, or conformation space, for example, but they could also include problem parameters such as boundary conditions or material parameters as described later. We seek a solution for $(x_1, \ldots, x_D) \in \Omega_1 \times \cdots \times \Omega_D$.

The PGD yields an approximate solution in the separated form:

$$u(x_1, \ldots, x_D) \approx \sum_{i=1}^{N} F_i^1(x_1) \cdot \ldots \cdot F_i^d(x_D). \tag{1.13}$$

The PGD approximation is thus a sum of N functional products involving each a number D of functions $F_i^j(x_j)$ that are unknown *a priori*. It is constructed by successive enrichment, whereby each functional product is determined in sequence. At a particular enrichment step $n+1$, the functions $F_i^j(x_j)$ are known for $i \leq n$ from the previous steps, and one must compute the new product involving the D unknown functions $F_{n+1}^j(x_j)$. This is achieved by invoking the weak form of the problem under consideration. The resulting discrete system is nonlinear, which implies that iterations are needed at each enrichment step. A low-dimensional problem can thus be defined in Ω_j for each of the D functions $F_{n+1}^j(x_j)$. The interested reader can refer to the primer [19] for a valuable introduction on the PGD technology.

If M nodes are used to discretize each coordinate, the total number of PGD unknowns is $N \cdot M \cdot D$ instead of the M^D degrees of freedom involved in standard mesh-based discretizations.

1.4 Revisiting the Simulation Challenges

In this section we describe the way in which separated representation can be used for circumventing or alleviating the challenges enumerated in Sect. 1.1.

1.4.1 Multidimensional Models

We consider the model

$$\mathcal{L}(u) = f(x_1, \cdots, x_D) \tag{1.14}$$

defined in $\Omega_1 \times \cdots \times \Omega_D$ with $u(x_1, \cdots, x_D)$. In the previous equation $\mathcal{L}$ represents the differential operator involving the D coordinates that for the sake of simplicity, and without loss of generality, is assumed linear and accepting the following decomposition

$$\mathcal{L}(u) = \sum_{i=1}^{i=D} \mathcal{G}_i(x_1, \cdots, x_D) \cdot \mathcal{L}_{x_i}(u) \tag{1.15}$$

where $\mathcal{L}_{x_i}$ denotes a linear differential operator with respect to the x_i coordinate.

The solution procedure consists of assuming the separated representation

$$u(x_1, \cdots, x_D) \approx \sum_{i=1}^{N} F_i^1(x_1) \cdot \ldots \cdot F_i^D(x_D). \tag{1.16}$$

and insert it in the weak form associated with Eq. (1.14)

$$\int_{\Omega_1 \times \cdots \times \Omega_D} u^* \cdot (\mathcal{L}(u) - f) \ dx_1 \ \cdots \ dx_D = 0 \tag{1.17}$$

As just described the solution is constructed by successive enrichments, whereby each functional product is determined in sequence. At a particular enrichment step n, the functions $F_i^j(x_j)$ are known for $i < n$ from the previous steps, and one must compute the new product involving the D unknown functions $F_n^j(x_j)$.

As the problem of calculating the D functions $F_n^j(x_j)$ at enrichment step n is nonlinear, the use of an appropriate linearization scheme is mandatory. The simplest strategy widely considered in our works consists of using an alternated direction fixed point algorithm in which $F_n^j(x_j)$ is computed assuming all the other functions $F_n^k(x_k), \forall k \neq j$ known. In this case the test function u^* is written

$$u^*(x_1, \cdots, x_D) = (F_n^j)^* \cdot \prod_{k=1, k \neq j}^{D} F_n^k(x_k) \tag{1.18}$$

By introducing Eqs. (1.16) and (1.18) into (1.17) and integrating in $\Omega_1 \times \cdots \times \Omega_{j-1} \times \Omega_{j+1} \times \cdots \times \Omega_D$ it results that

$$\int_{\Omega_j} (F_n^j)^* \cdot \left(\alpha^j \cdot \mathscr{L}_{x_j}(F_n^j(x_j)) + \beta^j \cdot F_n^j + g_n^j(x_j) \right) dx_j = 0 \qquad (1.19)$$

where α^j and β^j are two constants and $g_n^j(x_j)$ is the function that results from the integration of $f(x_1, \cdots, x_D)$ and $\mathscr{L}(\sum_{i=1}^{n-1} F_i^1(x_1) \cdot \ldots \cdot F_i^d(x_D))$.

1.4.2 Efficient Transient Non-incremental Solutions

As just discussed standard incremental transient solutions require in general the solution of a 3D problem at each time step. When the number of time steps and/or the size of the space problems to be solved at each time step increase, standard incremental solutions become rapidly inefficient. POD-based model order reduction can alleviate such drawback by considering a reduced space approximation basis that allows approximating the model solution at each time step. Thus the solution procedure remains incremental but now a small linear system of equations must be solved at each time step.

The main difficulty when considering such an approach lies in the fact that despite the fact of using a reduced basis, in general, one must evaluate the whole discrete matrix $\mathbf{G}^m$ at each time step t_m (1.8) and then by pre and post-multiplying it by $\mathbf{B}^T$ and $\mathbf{B}$ respectively one obtains the reduced linear system (1.11) to be solved at each time step t_m. While standard integration procedures spend most of the computing time in the solution of large linear systems of size M, one at each time step, POD-based reduced models are constrained by the necessity of assembling the whole system prior to solving the resulting reduced linear system.

Many possibilities exist to alleviate this task. Ryckelynck proposed the so-called hyper-reduction framework [20] that consists in constructing a few rows of matrix $\mathbf{G}^m$. In fact as the reduced model only involves N unknowns, N equations should be enough. Even if in that case $\mathbf{G}^m$ is singular, it is not after pre and post-multiplying it by $\mathbf{B}^T$ and $\mathbf{B}$ respectively because the functions involved in the reduced basis, i.e., the rows of matrix $\mathbf{B}$, are defined in the whole space domain Ω.

The trickiest point within the hyper-reduction framework lies in the choice of these N equations. Different variants have been proposed and in general they employ slightly more equations that strictly needed in order to enhance accuracy. Other techniques (Gappy-POD, GNAT, Missing Point Estimation, ...) consider alternative strategies for the construction of the reduced discrete matrix. These techniques are not addressed in the present work but there exist a vast literature on all them.

Obviously when addressing linear models in fixed domains with constant model parameters, matrix $\mathbf{G}$ remains unchanged during the time evolution. In this case matrix $\mathbf{G}$ could be calculated and factorized only once. Thus, in this particular scenario the interest in using model order reduction is not plenty justified.

The nonlinear case is the main domain of interest for considering model reduction because except when using very particular nonlinear solvers, like the asymptotic numerical method [21], one should construct and factorize the discrete matrix $\mathbf{G}^m$ at least one time at each time step t_m. In this context interpolation techniques applied on the nonlinear term allows impressive computing time savings [22, 23].

Other alternative route consists of making use of a non-incremental integration scheme based on the space-time separated representation. For illustrating it, we consider the simplest equation in which using the notation introduced in the previous section reads:

$$\mathscr{L}(u) = f(\mathbf{x}, t) \tag{1.20}$$

defined in $\Omega \times \mathscr{I}$ with $u(\mathbf{x}, t)$. We consider, again without loss of generality, the simplest case

$$\mathscr{L}(u) = \mathscr{L}_t(u) + \mathscr{L}_x(u) \tag{1.21}$$

where $\mathscr{L}_t = \frac{\partial}{\partial t}$ and $\mathscr{L}_x$ is the Laplace operator, i.e., $\mathscr{L}_x = \nabla^2$.

The solution procedure consists of assuming the space-time separated representation

$$u(\mathbf{x}, t) \approx \sum_{i=1}^{N} X_i(\mathbf{x}) \cdot T_i(t) \tag{1.22}$$

and insert it in the weak form associated with Eq. (1.20)

$$\int_{\Omega \times \mathscr{I}} u^* \cdot (\mathscr{L}(u) - f)\; d\mathbf{x}\, dt = 0. \tag{1.23}$$

At a particular enrichment step n, functions $X_i(\mathbf{x})$ and $T_i(t)$, $i < n$, are known from the previous steps, and one must compute the new term involving functions $X_n(\mathbf{x})$ and $T_n(t)$.

As the problem of calculating functions X_n and T_n at enrichment step n is non-linear, the use of an appropriate linearization scheme is mandatory. The simplest consists of using an alternated direction fixed point algorithm in which X_n is computed assuming T_n known, and then updating T_n from the just calculated X_n. The iteration procedure continues until reaching convergence (the fixed point).

When looking for X_n the test function u^* is written

$$u^*(\mathbf{x}, t) = X_n^* \cdot T_n. \tag{1.24}$$

By introducing Eqs. (1.24) and (1.22) into (1.23) and integrating in $\mathscr{I}$ it results that

$$\int_{\Omega} X_n^* \cdot \left(\alpha^x \cdot \mathscr{L}_x(X_n(\mathbf{x})) + \beta^x \cdot X_n(\mathbf{x}) + g_n^x(\mathbf{x})\right)\, d\mathbf{x} = 0 \tag{1.25}$$

where α^x and β^x are two constants and $g_n^x(\mathbf{x})$ is the function that results from the integration in $\mathscr{I}$ of $f(\mathbf{x}, t)$ and $\mathscr{L}(\sum_{i=1}^{n-1} X_i(\mathbf{x}) \cdot T_i(t))$.

The weak form (1.25) can be solved by using any appropriate discretization technique for elliptic boundary value problems.

Now, when looking for T_n the test function u^* writes

$$u^*(\mathbf{x}, t) = X_n \cdot T_n^*. \tag{1.26}$$

By introducing Eqs. (1.26) and (1.22) into (1.23) and integrating in Ω it results

$$\int_{\mathscr{I}} T_n^* \cdot \left(\alpha^t \cdot \mathscr{L}_t(T_n(t)) + \beta^t \cdot T_n(t) + g_n^t(t)\right) \, dt = 0 \tag{1.27}$$

where α^t and β^t are two constants and $g_n^t(t)$ is the function that results from the integration in Ω of $f(\mathbf{x}, t)$ and $\mathscr{L}(\sum_{i=1}^{n-1} X_i(\mathbf{x}) \cdot T_i(t))$.

This equation can be solved by using any appropriate discretization technique for initial value problems.

It can be noticed that when using the non-incremental space-time separated representation instead of solving a 3D problem at each time step, we must solve the order of N 3D problems related to the calculation of functions $X_i(\mathbf{x})$ and N one-dimensional problems for calculating function $T_i(t)$. Even when using an extremely small time step, the computing time related to the solution of the 1D problems associated with functions $T_i(t)$ is negligible with respect to the computing time required for solving the 3D problems associated with $X_i(\mathbf{x})$. Thus finally the complexity does not depend on the time step when using a space-time separated representation and then the computing time savings can be impressive, of several orders of magnitude in many cases.

1.4.3 Space Separation in Degenerated Domains

In what follows we consider a problem defined in a plate domain in which we are interested in calculating its 3D solution. In many cases when models are defined in plate or shell like domains different simplifying hypotheses can be introduced in order to reduce its 3D complexity to 2D. It is under this rationale that plate and shell elastic theories were developed in structural mechanics and lubrication models where proposed for addressing viscous flows. However in many applications the complex physics involved do not allow for such dimensional reduction and the only valuable route consists in solving the full 3D model. However, in such degenerated domains 3D meshes involve too many elements with the associated impact in the solution procedure efficiency.

We define the 3D plate domain Ω from $\Omega = \Xi \times \Gamma$, where Γ represents the plate thickness. Thus $(x, y, z) \in \Omega$ can be expressed from $(\mathbf{x}, z)$ with $\mathbf{x} \in \Xi \subset \mathscr{R}^2$ and $z \in \Gamma \subset \mathscr{R}$.

We consider again for the sake of simplicity the 3D heat equation defined in $\Omega = \Xi \times \Gamma$,

$$\mathscr{L}(u) = f(\mathbf{x}, z), \tag{1.28}$$

with $u(\mathbf{x}, z)$. We consider, again without loss of generality, the simplest case

$$\mathscr{L}(u) = \mathscr{L}_x(u) + \mathscr{L}_z(u), \tag{1.29}$$

where $\mathscr{L}_x = \frac{\partial^2}{\partial x^2} + \frac{\partial^2}{\partial y^2}$ and $\mathscr{L}_z = \frac{\partial^2}{\partial z^2}$.

The solution procedure consists in assuming an in-plane-out-of-plane separated representation

$$u(\mathbf{x}, z) \approx \sum_{i=1}^{N} X_i(\mathbf{x}) \cdot Z_i(z), \tag{1.30}$$

and insert it in the weak form associated with Eq. (1.28)

$$\int_{\Xi \times \Gamma} u^* \cdot (\mathscr{L}(u) - f) \, d\mathbf{x} \, dz = 0. \tag{1.31}$$

At a particular enrichment step n, functions $X_i(\mathbf{x})$ and $Z_i(z)$, $i < n$, are known from the previous steps, and one must compute the new term involving functions $X_n(\mathbf{x})$ and $Z_n(z)$.

As the problem of calculating functions X_n and Z_n at enrichment step n is nonlinear, the use of an appropriate linearization scheme is mandatory. Again, the simplest consists in using an alternated direction fixed point algorithm in which X_n is computed assuming Z_n known, and then updating Z_n from the just calculated X_n. The iteration procedure continues until reaching convergence (the fixed point).

When looking for X_n the test function u^* is written

$$u^*(\mathbf{x}, z) = X_n^* \cdot Z_n. \tag{1.32}$$

By introducing Eqs. (1.32) and (1.30) into (1.31) and integrating in Γ it results that

$$\int_{\Xi} X_n^* \cdot \left(\alpha^x \cdot \mathscr{L}_x(X_n(\mathbf{x})) + \beta^x \cdot X_n(\mathbf{x}) + g_n^x(\mathbf{x})\right) \, d\mathbf{x} = 0 \tag{1.33}$$

where α^x and β^x are two constants and $g_n^x(\mathbf{x})$ is the function that results from the integration in Γ of $f(\mathbf{x}, t)$ and $\mathscr{L}(\sum_{i=1}^{n-1} X_i(\mathbf{x}) \cdot Z_i(z))$.

The weak form (1.33) can be solved by using any appropriate discretization technique for elliptic boundary value problems.

Now, when looking for Z_n the test function u^* is written

$$u^*(\mathbf{x}, t) = X_n \cdot Z_n^*. \tag{1.34}$$

By introducing Eqs. (1.34) and (1.30) into (1.31) and integrating in Ξ it results

$$\int_{\Gamma} Z_n^* \cdot \left(\alpha^z \cdot \mathscr{L}_z(Z_n(z)) + \beta^z \cdot Z_n(z) + g_n^z(z) \right) \, dz = 0, \tag{1.35}$$

where α^z and β^z are two constants and $g_n^z(z)$ is the function that results from the integration in Ξ of $f(\mathbf{x}, z)$ and $\mathscr{L}(\sum_{i=1}^{n-1} X_i(\mathbf{x}) \cdot Z_i(z))$.

This equation can be solved by using any appropriate discretization of 1D boundary value problems.

It can be noticed that when using the in-plane-out-of-plane separated representation instead of solving a 3D problem, we must solved of the order of N 2D problems related to the calculation of functions $X_i(\mathbf{x})$ and N one-dimensional problems for calculating functions depending on the thickness coordinate z, $Z_i(z)$. Even when using a extremely fine discretization of the domain thickness, the computing time related to the solution of the 1D problems associated with functions $Z_i(z)$ is negligible with respected to the computing time required for solving the 2D problems associated with the calculation of functions $X_i(\mathbf{x})$. Thus finally the complexity does not depend on thickness representation when using in-plane-out-of-plane separated representations and then the computing time savings can be again impressive.

1.4.4 Parametric Solutions and Computational Vademecums

As emphasized in the first section of the present chapter, optimization, inverse identification and uncertainty quantification require the solution of many direct problems when considering standard procedures. In the same way real time simulation and the associated DDDAS are only possible if each solution can be performed fast enough.

There are many techniques that have been proposed in order to increase the solution efficiency. High performance computing and the recent GPUs based computations are contributing to enlarge the simulation capabilities. However, and despite impressive recent progress, many problems of industrial interest, in particular those related to decision making in complex design frameworks, remain unaffordable.

Within the separated representation framework the situation could be radically different. Due to the fact that separated representations can break the curse of dimensionality, one could envisage considering the parameters of interest (the ones involved in optimization—including geometrical parameters involved in shape optimization—or inverse identification, the ones having a stochastic distribution, or the ones concerned by real time simulation—boundary conditions, ...) as extra-coordinates. Then by solving only once and off-line the resulting multidimensional model we have access to the parametric solution that can be viewed as a sort of handbook, virtual chart or vademecum that can be then used on-line in real time and even in deployed platforms like tablets or smartphones. These vademecums can be viewed as a sort of modern nomograms related to complex multiparametric partial differential equations.

In this section, we consider again the heat transfer equation but in the present case parametrized by the thermal diffusivity k,

$$\mathscr{L}(u) = f(\mathbf{x}, t), \tag{1.36}$$

defined in $\Omega \times \mathscr{I} \times \mathscr{K}$ with $k \in \mathscr{K} \subset \mathscr{R}^+$ and $u(\mathbf{x}, t, k)$. We consider, again without loss of generality, the simplest case

$$\mathscr{L}(u) = \mathscr{L}_t(u) + k\ \mathscr{L}_x(u), \tag{1.37}$$

where $\mathscr{L}_t = \frac{\partial}{\partial t}$ and $\mathscr{L}_x$ is the Laplace operator, i.e., $\mathscr{L}_x = \nabla^2$.

The solution procedure consists in assuming the space-time-parameter separated representation

$$u(\mathbf{x}, t, k) \approx \sum_{i=1}^{N} X_i(\mathbf{x}) \cdot T_i(t) \cdot K_i(k), \tag{1.38}$$

and insert it in the weak form associated with Eq. (1.36)

$$\int_{\Omega \times \mathscr{I} \times \mathscr{K}} u^* \cdot (\mathscr{L}(u) - f)\ d\mathbf{x}\, dt\, dk = 0. \tag{1.39}$$

At a particular enrichment step n, functions $X_i(\mathbf{x})$, $T_i(t)$ and $K_i(k)$, $i < n$, are known from the previous steps, and one must compute the new term involving functions $X_n(\mathbf{x})$, $T_n(t)$ and $K_n(k)$.

As the problem of calculating functions X_n, T_n and K_n at enrichment step n is nonlinear, the use of an appropriate linearization scheme is mandatory. The simplest consists of using an alternated direction fixed point algorithm in which X_n is computed assuming T_n and K_n known, then updating T_n from the just calculated X_n and the old K_n and finally updating K_n from the just calculated X_n and T_n. The iteration procedure continues until reaching convergence, that is the fixed point.

When looking for X_n the test function u^* is written

$$u^*(\mathbf{x}, t, k) = X_n^* \cdot T_n \cdot K_n. \tag{1.40}$$

By introducing Eqs. (1.40) and (1.38) into (1.39) and integrating in $\mathscr{I} \times \mathscr{K}$ it results that

$$\int_{\Omega} X_n^* \cdot \left(\alpha^x \cdot \mathscr{L}_x(X_n(\mathbf{x})) + \beta^x \cdot X_n(\mathbf{x}) + g_n^x(\mathbf{x})\right) \, d\mathbf{x} = 0, \tag{1.41}$$

where α^x and β^x are two constants and $g_n^x(\mathbf{x})$ is the function that results from the integration in $\mathscr{I} \times \mathscr{K}$ of $f(\mathbf{x}, t)$ and $\mathscr{L}(\sum_{i=1}^{n-1} X_i(\mathbf{x}) \cdot T_i(t) \cdot K_i(k))$.

The weak form (1.41) can be solved by using any appropriate discretization technique for elliptic boundary value problems.

When looking for T_n the test function u^* is written

$$u^*(\mathbf{x}, t, k) = X_n \cdot T_n^* \cdot K_n. \tag{1.42}$$

By introducing Eqs. (1.42) and (1.38) into (1.39) and integrating in $\Omega \times \mathscr{K}$ it results

$$\int_{\mathscr{I}} T_n^* \cdot \left(\alpha^t \cdot \mathscr{L}_t(T_n(t)) + \beta^t \cdot T_n(t) + g_n^t(t)\right) \, dt = 0, \tag{1.43}$$

where α^t and β^t are two constants and $g_n^t(t)$ is the function that results from the integration in $\Omega \times \mathscr{K}$ of $f(\mathbf{x}, t)$ and $\mathscr{L}(\sum_{i=1}^{n-1} X_i(\mathbf{x}) \cdot T_i(t) \cdot K_i(k))$.

This equation can be solved by using any appropriate discretization technique for initial value problems.

Finally, when looking for K_n the test function u^* is written

$$u^*(\mathbf{x}, t, k) = X_n \cdot T_n \cdot K_n^*. \tag{1.44}$$

By introducing Eqs. (1.44) and (1.38) into (1.39) and integrating in $\Omega \times \mathscr{I}$ it results

$$\int_{\mathscr{K}} K_n^* \cdot \left(\alpha^k \cdot K_n(k) + \beta^k \cdot k \cdot K_n(k) + g_n^k(k)\right) \, dk = 0, \tag{1.45}$$

where α^k and β^k are two constants and $g_n^k(k)$ is the function that results from the integration in $\Omega \times \mathscr{I}$ of $f(\mathbf{x}, t)$ and $\mathscr{L}(\sum_{i=1}^{n-1} X_i(\mathbf{x}) \cdot T_i(t) \cdot K_i(k))$.

Equation (1.45) represents an algebraic problem, which is hardly a surprise since the original equation (1.36) does not contain derivatives with respect to the parameter k. Introduction of the parameter k as additional model coordinate does not increase the cost of a particular enrichment step. It does however necessitate more enrichment steps, i.e., more terms (higher N) in the decomposition (1.38).

In this case we can conclude that the complexity of the PGD procedure to compute the approximation (1.38) is of some tens of 3D steady-state problems (the cost related to the 1D IVP and algebraic problems being negligible with respect to the 3D problems). In a classical approach, one must solve for each particular value of the parameter k a 3D problem at each time step. In usual applications, this often implies the computation of several millions of 3D solutions. Clearly, the CPU time savings by applying the PGD can be of several orders of magnitude.

1.4.5 Discussion on the Coordinates Separation

When performing a separated representation of the solution of a problem described by a PDE the first step relies in the choice of the coordinates to be separated. We could consider the following scenarios:

- Efficient solvers for transient problems can be defined by applying a space-time separation:

$$u(\mathbf{x},t) \approx \sum_{i=1}^{N} X_i(\mathbf{x}) \cdot T_i(t). \tag{1.46}$$

  The constructor of that separated representation was illustrated in the previous section. We cited previously the pioneering works of Ladeveze's team in the field of structural mechanics. Space-time separated representations were also considered in the context of the multiscale coupling of diffusion and kinetic models endowed with very different characteristic times [24] as well as in the context of time multi-scale models [25].
- The fully three-dimensional solution of models defined in degenerate domains is also an appealing field of application of the PGD. Consider the unknown field $u(x, y, z)$ defined in a domain Ω. Two approaches come to mind:
  - Fully separated decomposition.

$$u(x,y,z) \approx \sum_{i=1}^{N} X_i(x) \cdot Y_i(y) \cdot Z_i(z). \tag{1.47}$$

    This strategy is particularly suitable for fully separable domains, i.e., $\Omega = \Omega_x \times \Omega_y \times \Omega_z$ [26, 27]. For general domains, embedding Ω into a larger separable domain $\tilde{\Omega} = \tilde{\Omega}_x \times \tilde{\Omega}_y \times \tilde{\Omega}_z$ can also be done, as described in [28].
  - 2D-1D decompositions are suitable in plate, shell or profiled geometries. In these cases the most natural decomposition reads

$$u(x,y,z) \approx \sum_{i=1}^{N} X_i(x,y) \cdot Z_i(z). \tag{1.48}$$

    In plate-type domains $z \in \mathscr{I}$, where $\mathscr{I}$ denotes the plate thickness dimension whereas in extruded profiles it denotes the extrusion direction being Ξ the transversal section.

- Finally, for applications requiring many solutions of a particular model, it suffices to introduce all sources of variability as extra-coordinates. The solution of the resulting parametric multidimensional model is then sought in the separated form

$$u(\mathbf{x}, t, p_1, \cdots p_Q) \approx \sum_{i=1}^{N} X_i(\mathbf{x}) \cdot T_i(t) \cdot P_i^1(p_1) \cdots P_i^Q(p_Q), \qquad (1.49)$$

where the p_i denote the different problem parameters such as material parameters, boundary conditions, applied loads, initial conditions, and/or geometrical parameters (see [29] and [30] and the references therein).

1.5 A Brief State of the Art on PGD-Based Model Order Reduction

1.5.1 Multidimensional Models

As discussed previously, some models are inherently defined in a configurational or conformation space that is high-dimensional in nature. Its discretization leads to the well-known curse of dimensionality if traditional, mesh-based techniques are applied.

Separated representations were applied for solving the multidimensional Fokker-Planck equation describing complex fluids in the kinetic theory framework. In [31], the authors addressed the solution of the linear Fokker-Planck equation describing multi-bead-spring molecular models in the steady state and in homogeneous flows. In that case the distribution function describing the molecular conformation only depends on the molecules' configurational coordinates $\mathbf{q}_1, \cdots \mathbf{q}_D$ (see Fig. 1.1). The molecular conformation distribution function $\Psi(\mathbf{q}_1, \cdots, \mathbf{q}_D)$ results from the solution of the associated balance Fokker-Planck equation:

$$\frac{\partial \Psi}{\partial t} = -\sum_{i=1}^{i=D} \left(\dot{\mathbf{q}}_i \cdot \Psi\right), \qquad (1.50)$$

where the conformational velocities $\dot{\mathbf{q}}_i$ depend on the molecule conformation and the molecule-fluid interactions. The solution was searched in the separated form:

$$\Psi(\mathbf{q}_1, \cdots, \mathbf{q}_D) \approx \sum_{i=1}^{i=N} F_i^1(\mathbf{q}_1) \times \cdots \times F_i^D(\mathbf{q}_D). \qquad (1.51)$$

The solution procedure was extended to nonlinear kinetic theory descriptions of more complex molecular descriptions in [32]. The transient solution was addressed in [33] in which the time was added as an extra-coordinate. Transient solutions making use of reduced bases in the context of an adaptive proper orthogonal decomposition [7] were considered in the case of low dimensional configuration spaces: the FENE model was addressed in [15] and liquid crystalline polymers in [34]. A deeper analysis

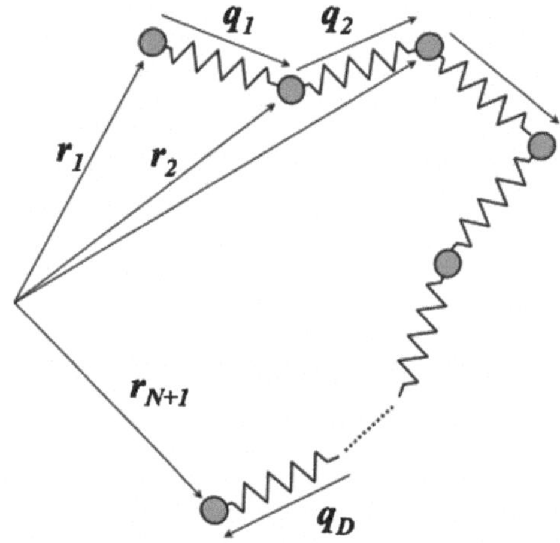

Fig. 1.1 Molecular conformation

of nonlinear and transient models was considered in [35]. Complex fluid models were coupled with complex flows in [36] and [37] opening very encouraging perspectives and claiming the necessity of defining efficient stabilizations. Finally, in [38] the PGD was applied for solving the stochastic equation within the Brownian Configuration Field framework. The interested reader can refer to [30, 39] and the references therein for an exhaustive overview of PGD in computational rheology.

Multidimensional models encountered in the finer descriptions of matter (ranging from quantum chemistry to statistical mechanics descriptions) were revisited in [40]. The multidimensional chemical master equation was efficiently solved in [3, 41]. Langer's equation governing phase transitions was considered in [42].

1.5.2 Separating the Physical Space

Other models are not inherently defined in a high-dimensional space, but can nevertheless be treated efficiently in a separated manner. Models defined in cubic domains suggest the following separated representation,

$$u(x, y, z) \approx \sum_{i=1}^{i=N} X_i(x) \cdot Y_i(y) \cdot Z_i(z) \tag{1.52}$$

Thus, the $3D$ solution results in a sequence of $1D$ solutions for computing functions $X_i(x)$, $Y_i(y)$ and $Z_i(z)$. This kind of domains are very frequent when addressing homogenization problems in which the elementary volume element is a cube. The interested reader can refer to [26] or [43]. Fully separated representations in complex, non-hypercubic domains can be equally performed [28]. In that work the enforcement of non-homogeneous Dirichlet boundary conditions was also analyzed, giving rise to a generalized form of the PGD approach.

In the case of plates, shells or extruded geometries, one could consider also advantageously the separated approximation [44–47]:

$$u(x, y, z) \approx \sum_{i=1}^{i=N} X_i(x, y) \cdot Z_i(z) \tag{1.53}$$

In the case of models defined in plate geometries (x, y) represent the in-plane coordinates and z the thickness direction. In the case of extruded profiles (x, y) represents the surface extruded in the z direction. This kind of representation makes possible fully $3D$ solutions with numerical complexities (and therefore computation times) characteristic of $2D$ solutions, without any simplifying kinematic assumption.

1.5.3 Parametric Models: A Route to Efficient Optimization, Inverse Identification and Real Time Simulation

Usual computational mechanics models could be enriched by introducing several extra-coordinates. Thus, adding some new coordinates to models initially non high-dimensional, could lead to new, never "before" explored insights in the physics as previously illustrated in the context of a parametric thermal model.

This kind of parametric modeling was addressed in [44, 48, 49] where material parameters were introduced as extra-coordinates. In [50] and [27], thermal conductivities, macroscopic temperature and its time evolution were introduced as extra-coordinates for computing linear and nonlinear homogenization. In [44], the anisotropy directions of plies involved in a composite laminate were considered as extra-coordinates. By assuming a certain uncertainty in the real orientation of such plies, authors evaluated the envelope of the resulting distorted structures due to the thermomechanical coupling.

Moreover, as soon as the separated representation of the parametric solution has been computed off-line, its on-line use only needs to particularize such solutions for a desired set of parameters. Obviously, this task can be performed very fast, many times in real time, and by using light computing platforms, as smartphones or tablets.

The interested reader can refer to [51] and the references therein for an exhaustive review on PGD-based parametric solutions.

1.5.4 Real-Time Simulation, DDDAS and More

It is easy to understand that after performing this type of calculations, in which parameters are considered advantageously as new coordinates of the model, a posteriori inverse identification or optimization can be easily handled. This new PGD framework allows us to perform this type of calculations very efficiently, because in fact all possible solutions have been previously computed in the form of a separated, high-dimensional solution so that they constitute a simple post-processing of this general solution.

Process optimization was considered in [52], for instance. Shape optimization was performed by considering all the geometrical parameters as extra-coordinates, leading to the model solution in any of the geometries generated by the parameters considered as extra-coordinates [53]. More complex scenarios were considered in [54] where the Laplace equation was solved in a parametrized domain. This strategy could be an alternative to the POD-based shape optimization considered in [55].

Inverse methods in the context of real-time simulations were addressed in [56] and were coupled with control strategies in [57] as a first step towards DDDAS (dynamic data-driven application systems). Moreover, because the general parametric solution was pre-computed off-line, it can be used on-line under real time constraints and using light computing platforms like smartphones [44, 57], that constitutes a first step towards the use of this kind of information in augmented reality platforms.

As mentioned before, surgical simulators must operate at frequencies higher than 500 Hz. The use of model reduction seems to be an appealing alternative for reaching such performances. However, techniques based on the use of the POD, PODI, even combined with an asymptotic numerical methods to avoid the computation of the tangent matrix, exhibit serious difficulties to fulfill such requirements as discussed in [58–61].

1.5.5 Miscellaneous

There is a variety of open questions when applying the PGD-based discretization techniques. Stabilization is one of the main issues, and concerns both convective and mixed formulations stabilizations. In [62] standard convective stabilization approaches were generalized in the separated representations framework.

Another important question lies in the error estimation. A first attempt was considered in [63]. Coupling standard mesh based discretization techniques with reduced bases (POD and PGD) is an issue of major interest in order to represent localized behaviors (e.g., discontinuities, singularities, boundary layers, ...). Some coupling strategies were proposed in [64] and [25].

Multi-scale and multi-physics nonlinear coupled models involving different characteristic times were efficiently coupled in [24] where a globalization of local problems to speed-up the simulation in a separated representation framework has been proposed. Later, in [65] the time axis was transformed into a two-dimensional domain to account for two very different time scales. Parallel time integration strategies, inspired by the ones proposed in [26] were considered in [66] within the PGD framework. The coupling of other multi-physic models in the context of composites manufacturing processes was performed in [67]. Non-incremental parametrical solutions defined in evolving domains were addressed in [68]. In [69] the solution of the electromagnetic equations was considered.

In [70] the PGD was introduced in the boundary element method framework for solving transient models where few works concerning the use of reduced bases exist [13]. Because the use of a space-time separated representation the time-dependent kernel is no longer needed, and only the steady state kernel functions are needed. The space-time separation implies significant computational time savings.

Finally, in [71] we considered a PGD approach to augmented learning in science and engineering higher education.

References

1. E. Cancès, M. Defranceschi, W. Kutzelnigg, C. Le Bris, Y. Maday, Computational quantum chemistry: a primer, in *Handbook of Numerical Analysis*, vol. X (Elsevier, Amsterdam, 2003) pp. 3–270, 2003.
2. B.B. Bird, C.F. Curtiss, R.C. Armstrong, O. Hassager, Dynamics of polymeric liquids, in: *Kinetic Theory*, vol. 2 (Wiley, New York, 1987)
3. A. Ammar, E. Cueto, F. Chinesta, Reduction of the chemical master equation for gene regulatory networks using proper generalized decompositions. Int. J. Numer. Meth. Biomed. Eng **28**(9), 960–973 (2012)
4. NSF Final Report, *DDDAS Workshop 2006* (Arlington, VA, USA, 2006)
5. F. Darema, Engineering/Scientific and commercial applications: differences, similarities, and future evolution, in *Proceedings of the Second Hellenic European Conference on Mathematics and Informatics*. HERMIS, vol 1, (1994) pp. 367–374
6. J.T. Oden, T. Belytschko, J. Fish, T.J.R. Hughes, C. Johnson, D. Keyes, A. Laub, L. Petzold, D. Srolovitz, S.Yip, Simulation-based engineering science: revolutionizing engineering science through simulation. NSF Blue Ribbon Panel on SBES (2006)
7. D. Ryckelynck, F. Chinesta, E. Cueto, A. Ammar, On the a priori model reduction: overview and recent developments. State of the art reviews. Arch. Comput. Meth. Eng. **13**(1), 91–128 (2006)
8. R.A. Bialecki, A.J. Kassab, A. Fic, Proper orthogonal decomposition and modal analysis for acceleration of transient FEM thermal analysis. Int. J. Numer. Meth. Eng. **62**, 774–797 (2005)
9. J. Burkardt, M. Gunzburger, H-Ch. Lee, POD and CVT-based reduced-order modeling of Navier-Stokes flows. Comput. Meth. Appl. Mech. Eng. **196**, 337–355 (2006)
10. M.D. Gunzburger, J.S. Peterson, J.N. Shadid, Reduced-order modeling of time-dependent PDEs with multiple parameters in the boundary data. Comput. Meth. Appl. Mech. Eng. **196**, 1030–1047 (2007)
11. Y. Maday, E.M. Ronquist, The reduced basis element method: application to a thermal fin problem. SIAM J. Sci. Comput. **26**(1), 240–258 (2004)
12. H.M. Park, D.H. Cho, The use of the Karhunen-Loeve decomposition for the modelling of distributed parameter systems. Chem. Eng. Sci **51**, 81–98 (1996)
13. D. Ryckelynck, L. Hermanns, F. Chinesta, E. Alarcon, An efficient a priori model reduction for boundary element models. Eng. Anal. Boundary Elem. **29**, 796–801 (2005)
14. D. Ryckelynck, A priori hyper-reduction method: an adaptive approach. J. Comput. Phys. **202**, 346–366 (2005)
15. A. Ammar, D. Ryckelynck, F. Chinesta, R. Keunings, On the reduction of kinetic theory models related to finitely extensible dumbbells. J. Nonnewton. Fluid Mech. **134**, 136–147 (2006)
16. P. Ladeveze, *Nonlinear Computational Structural Mechanics* (Springer, NY, 1999)

17. P. Ladevèze, J.-C. Passieux, D. Néron, The latin multiscale computational method and the proper generalized decomposition. Comput. Methods Appl. Mech. Eng. **199**(21–22), 1287–1296 (2010)
18. A. Nouy, P. Ladeveze, Multiscale computational strategy with time and space homogenization: a radial-type approximation technique for solving microproblems, Int. J. Multiscale Comput. Eng. **170**(2)557–574 (2004)
19. F. Chinesta, R. Keunings, A. Leygue, *The Proper Generalized Decomposition for Advanced Numerical Simulations. A Primer (Springerbriefs)* (Springer, New York, 2014)
20. D. Ryckelynck, Hyper-reduction of mechanical models involving internal variables. Int. J. Numer. Meth. Eng. **77**(1), 75–89 (2008)
21. B. Cochelin, N. Damil, M. Potier-Ferry, The asymptotic numerical method: an efficient perturbation technique for nonlinear structural mechanics. Revue Europeenne des Elements Finis **3**, 281–297 (1994)
22. M. Barrault, Y. Maday, N.C. Nguyen, A.T. Patera, An "empirical interpolation" method: application to efficient reduced-basis discretization of partial differential equations. Comptes Rendus Mathematique **339**(9), 667–672 (2004)
23. S. Chaturantabut, D.C. Sorensen, Nonlinear model reduction via discrete empirical interpolation. SIAM J. Sci. Comput. **32**, 2737–2764 (2010)
24. F. Chinesta, A. Ammar, E. Cueto, Proper generalized decomposition of multiscale models. Int. J. Numer. Meth. Eng. **83**(8–9), 1114–1132 (2010)
25. A. Ammar, F. Chinesta, E. Cueto, Coupling finite elements and proper generalized decomposition. Int. J. Multiscale Comput. Eng. **9**(1), 17–33 (2011)
26. F. Chinesta, A. Ammar, F. Lemarchand, P. Beauchene, F. Boust, Alleviating mesh constraints: model reduction, parallel time integration and high resolution homogenization. Comput. Methods Appl. Mech. Eng. **197**(5), 400–413 (2008)
27. H. Lamari, A. Ammar, P. Cartraud, G. Legrain, F. Jacquemin, F. Chinesta, Routes for efficient computational homogenization of non-linear materials using the proper generalized decomposition. Arch. Comput. Meth. Eng. **17**(4), 373–391 (2010)
28. D. Gonzalez, A. Ammar, F. Chinesta, E. Cueto, Recent advances in the use of separated representations. Int. J. Numer. Meth. Eng. **81**(5), 637–659 (2010)
29. F. Chinesta, A. Ammar, E. Cueto, Recent advances and new challenges in the use of the proper generalized decomposition for solving multidimensional models. Arch. Comput. Meth. Eng. **17**(4), 327–350 (2010)
30. F. Chinesta, A. Ammar, A. Leygue, R. Keunings, An overview of the proper generalized decomposition with applications in computational rheology. J. Nonnewton. Fluid Mech. **166**, 578–592 (2011)
31. A. Ammar, B. Mokdad, F. Chinesta, R. Keunings, A new family of solvers for some classes of multidimensional partial differential equations encountered in kinetic theory modeling of complex fluids. J. Nonnewton. Fluid Mech. **139**, 153–176 (2006)
32. B. Mokdad, E. Pruliere, A. Ammar, F. Chinesta, On the simulation of kinetic theory models of complex fluids using the Fokker-Planck approach. Appl. Rheol. **17**(2), 1–14, 26494 (2007)
33. A. Ammar, B. Mokdad, F. Chinesta, R. Keunings, A new family of solvers for some classes of multidimensional partial differential equations encountered in kinetic theory modeling of complex fluids (Part II). J. Nonnewton. Fluid Mech. **144**, 98–121 (2007)
34. A. Ammar, E. Pruliere, F. Chinesta, M. Laso, Reduced numerical modeling of flows involving liquid-crystalline polymeres. J. Nonnewton. Fluid Mech. **160**, 140–156 (2009)
35. A. Ammar, M. Normandin, F. Daim, D. Gonzalez, E. Cueto, F. Chinesta, Non-incremental strategies based on separated representations: applications in computational rheology. Commun. Math.Sci **8**(3), 671–695 (2010)
36. E. Pruliere, A. Ammar, N. El Kissi, F. Chinesta, Recirculating flows involving short fiber suspensions: numerical difficulties and efficient advanced micro-macro solvers, State of the art reviews. Arch. Comput. Meth. Eng **16**, 1–30 (2009)
37. B. Mokdad, A. Ammar, M. Normandin, F. Chinesta, J.R. Clermont, A fully deterministic micro-macro simulation of complex flows involving reversible network fluid models. Math. Comput. Simul. **80**, 1936–1961 (2010)

38. F. Chinesta, A. Ammar, A. Falco, M. Laso, On the reduction of stochastic kinetic theory models of complex fluids. Model. Simul. Mater. Sci. Eng. **15**, 639–652 (2007)
39. F. Chinesta, P. Ladeveze, E. Cueto, A short review in model order reduction based on proper generalized decomposition. Arch. Comput. Meth. Eng. **18**, 395–404 (2011)
40. A. Ammar, F. Chinesta, P. Joyot, The nanometric and micrometric scales of the structure and mechanics of materials revisited: an introduction to the challenges of fully deterministic numerical descriptions. Int. J. Multiscale Comput. Eng. **6**(3), 191–213 (2008)
41. F. Chinesta, A. Ammar, E. Cueto, On the use of proper generalized decompositions for solving the multidimensional chemical master equation. Eur. J. Comput. Mech. **19**, 53–64 (2010)
42. H. Lamari, A. Ammar, A. Leygue, F. Chinesta, On the solution of the multidimensional Langer's equation by using the proper generalized decomposition method for modeling phase transitions. Model. Simul. Mater. Sci. Eng. **20**, 015007 (2012)
43. E. Pruliere, F. Chinesta, A. Ammar, A. Leygue, A. Poitou, On the solution of the heat equation in very thin tapes. Int. J. Therm. Sci. **65**, 148–157 (2013)
44. B. Bognet, A. Leygue, F. Chinesta, A. Poitou, F. Bordeu, Advanced simulation of models defined in plate geometries: 3D solutions with 2D computational complexity. Comput. Methods Appl. Mech. Eng. **201**, 1–12 (2012)
45. B. Bognet, A. Leygue, F. Chinesta, On the fully 3D simulation of thermoelastic models defined in plate geometries. Eur. J. Comput. Mech. **21**(1–2), 40–51 (2012)
46. A. Leygue, F. Chinesta, M. Beringhier, T.L. Nguyen, J.C. Grandidier, F. Pasavento, B. Schrefler, Towards a framework for non-linear thermal models in shell domains. Int. J. Numer. Meth. Heat Fluid Flow **23**(1), 55–73 (2013)
47. B. Bognet, A. Leygue, F. Chinesta, Separated representations of 3D elastic solutions in shell geometries. Adv. Model. Simul. Mater. Sci. Eng. (in press) **1**(1), 1–34 (2014)
48. A. Ammar, M. Normandin, F. Chinesta, Solving parametric complex fluids models in rheometric flows. J. Nonnewton. Fluid Mech. **165**, 1588–1601 (2010)
49. E. Pruliere, F. Chinesta, A. Ammar, On the deterministic solution of multidimensional parametric models by using the proper generalized decomposition. Math. Comput. Simul. **81**, 791–810 (2010)
50. H. Lamari, F. Chinesta, A. Ammar, E. Cueto, Non-conventional numerical strategies in the advanced simulation of materials and processes. Int. J. Mod. Manuf. Technol. **1**, 49–56 (2009)
51. F. Chinesta, A. Leygue, F. Bordeu, J.V. Aguado, E. Cueto, D. Gonzalez, I. Alfaro, A. Ammar, A. Huerta, Parametric PGD based computational vademecum for efficient design, optimization and control. Arch. Comput. Meth. Eng. **20**(1), 31–59 (2013)
52. Ch. Ghnatios, F. Chinesta, E. Cueto, A. Leygue, P. Breitkopf, P. Villon, Methodological approach to efficient modeling and optimization of thermal processes taking place in a die: application to pultrusion. Compos. A **42**, 1169–1178 (2011)
53. A. Leygue, E. Verron, A first step towards the use of proper general decomposition method for structural optimization. Arch. Comput. Meth. Eng. **17**(4), 465–472 (2010)
54. A. Ammar, A. Huerta, A. Leygue, F. Chinesta, E. Cueto, Parametric solutions involving geometry: a step towards efficient shape optimization. Comput. Methods Appl. Mech. Eng. **268**, 178–193 (2014)
55. F. Schmidt, N. Pirc, M. Mongeau, F. Chinesta, Efficient mould cooling optimization by using model reduction. Int. J. Mater. Form. **4**(1), 71–82 (2011)
56. D. Gonzalez, F. Masson, F. Poulhaon, A. Leygue, E. Cueto, F. Chinesta, Proper generalized decomposition based dynamic data-driven inverse identification. Math. Comput. Simul. **82**(9), 1677–1695 (2012)
57. Ch. Ghnatios, F. Masson, A. Huerta, E. Cueto, A. Leygue, F. Chinesta, Proper generalized decomposition based dynamic data-driven control of thermal processes. Comput. Methods Appl. Mech. Eng. **213**, 29–41 (2012)
58. S. Niroomandi, I. Alfaro, E. Cueto, F. Chinesta, Real-time deformable models of non-linear tissues by model reduction techniques. Comput. Methods Programs Biomed. **91**, 223–231 (2008)

59. S. Niroomandi, I. Alfaro, E. Cueto, F. Chinesta, Model order reduction for hyperelastic materials. Int. J. Numer. Meth. Eng. **81**(9), 1180–1206 (2010)
60. S. Niroomandi, I. Alfaro, E. Cueto, F. Chinesta, Accounting for large deformations in real-time simulations of soft tissues based on reduced order models. Comput. Methods Programs Biomed. **105**, 1–12 (2012)
61. S. Niroomandi, I. Alfaro, D. Gonzalez, E. Cueto, F. Chinesta, Real time simulation of surgery by reduced order modelling and X-FEM techniques. Int. J. Numer. Meth. Biomed. Eng **28**(5), 574–588 (2012)
62. D. Gonzalez, E. Cueto, F. Chinesta, P. Diez, A. Huerta, SUPG-based stabilization of proper generalized decompositions for high-dimensional advection-diffusion equations. Int. J. Numer. Meth. Eng. **94**(13), 1216–1232 (2013)
63. A. Ammar, F. Chinesta, P. Diez, A. Huerta, An error estimator for separated representations of highly multidimensional models. Comput. Methods Appl. Mech. Eng. **199**, 1872–1880 (2010)
64. A. Ammar, E. Pruliere, J. Ferec, F. Chinesta, E. Cueto, Coupling finite elements and reduced approximation bases. Eur. J. Comput. Mech. **18**(5–6), 445–463 (2009)
65. A. Ammar, F. Chinesta, E. Cueto, M. Doblare, Proper generalized decomposition of time-multiscale models. Int. J. Numer. Meth. Eng. **90**(5), 569–596 (2012)
66. F. Poulhaon, F. Chinesta, A. Leygue, A first step towards a PGD based parallelization strategy. Eur. J. Comput. Mech. **21**(3–6), 300–311 (2012)
67. E. Pruliere, J. Ferec, F. Chinesta, A. Ammar, An efficient reduced simulation of residual stresses in composites forming processes. Int. J. Mater. Form. **3**(2), 1339–1350 (2010)
68. A. Ammar, E. Cueto, F. Chinesta, Non-incremental PGD solution of parametric uncoupled models defined in evolving domains. Int. J. Numer. Meth. Eng. **93**(8), 887–904 (2013)
69. M. Pineda, F. Chinesta, J. Roger, M. Riera, J. Perez, F. Daim, Simulation of skin effect via separated representations. Int. J. Comput. Math. Electr. Electron. Eng. **29**(4), 919–929 (2010)
70. G. Bonithon, P. Joyot, F. Chinesta, P. Villon, Non-incremental boundary element discretization of parabolic models based on the use of proper generalized decompositions. Eng. Anal. Boundary Elem. **35**(1), 2–17 (2011)
71. C. Quesada, D. Gonzalez, I. Alfaro, F. Bordeu, A. Leygue, E. Cueto, A. Huerta, F. Chinesta. Real-time simulation on handheld devices for augmented learning in science and engineering. PLOS ONE (Submitted)

Chapter 2
Fine Description of Materials

[On quantum mechanics] I don't like it, and I'm sorry I ever had anything to do with it.

—Erwin Schroedinger

Nano-science and nano-technology, as well as fine modeling of the structure and mechanics of materials from nanometric to micrometric scales use descriptions ranging from the quantum to statistical mechanics. Here we revisit modeling at these scales, pointing out the main challenges related to the numerical solution of such models that sometimes are discrete but involve an extremely large number of particles (as it is the case of molecular dynamics simulations or coarse-grained molecular dynamics) and other times are continuous but they are defined in highly multidimensional spaces leading to the well known curse of dimensionality issues. The curse of dimensionality provoked by some of these deterministic models will be emphasized and their numerical implications will be addressed in the second part of this chapter within the PGD framework.

2.1 From Quantum Mechanics to Statistical Mechanics: A Walk on the Frontier of the Simulable World

2.1.1 The Finest Description: The Quantum Approach

The quantum state of a given electronic distribution could be determined by solving the Schrödinger equation. This equation for many years has been considered as one of the finest descriptions of the world. However, before focusing on the challenges of its numerical solution, we would like to recall that this equation is not relativistic so fails when to applied to describe heavy atoms. Moreover, the Pauli's principle

F. Chinesta and E. Cueto, *PGD-Based Modeling of Materials, Structures and Processes*, ESAFORM Bookseries on Material Forming, DOI: 10.1007/978-3-319-06182-5_2,

constraint was introduced in the Schrödinger formalism in an "ad hoc" way and constitutes, as we illustrate later, the main difficulty in its solution.

Some simplifying hypotheses are usually introduced, as for example the Born-Oppenheimer approximation that states that the nuclei can be in first approximation assumed as classical point-like particles, that the state of electrons only depends on the nuclei positions and that the electronic ground state corresponds to the one that minimizes the electronic energy for a nuclear configuration. The Schrödinger equation defines a multidimensional problem whose dimension increases linearly with the number of the electrons in the system.

Thus, the knowledge of a quantum system reduces to the determination of the wavefunction $\Psi(\mathbf{x}_1, \mathbf{x}_2, \ldots, \mathbf{x}_N, \; t; \mathbf{X}_1, \ldots, \mathbf{X}_M)$ (that establishes that the electronic wavefunction depends parametrically on the nuclei positions $\mathbf{X}_i$) whose evolution is governed by the Schrödinger equation:

$$i\hbar\frac{\partial \Psi}{\partial t} = -\frac{\hbar^2}{2m_e}\sum_{e=1}^{e=N} \nabla_e^2 \Psi + \sum_{e=1}^{e=N-1}\sum_{e'=e+1}^{e'=N} V_{ee'}\Psi + \sum_{e=1}^{e=N}\sum_{n=1}^{n=M} V_{en}\Psi \tag{2.1}$$

where N is the number of electrons and M the number of nuclei, the last ones assumed located and fixed at positions $\mathbf{X}_j$. Each electron is defined in the whole physical space $\mathbf{x}_j \in \mathbb{R}^3$, $i = \sqrt{-1}$, $\hbar$ represents the Planck's constant divided by 2π and m_e is the electron mass.

The differential operator ∇_e^2 is defined in the conformation space of each electron, i.e.: $\nabla_e^2 \equiv \frac{\partial^2}{\partial x_e^2} + \frac{\partial^2}{\partial y_e^2} + \frac{\partial^2}{\partial z_e^2}$. The Coulomb's potentials accounting for the electron-electron and electron-nuclei interactions are written:

$$V_{ee'} = \frac{(q_e)^2}{\|\mathbf{x}_e - \mathbf{x}_{e'}\|} \tag{2.2}$$

$$V_{en} = -\frac{q_n q_e}{\|\mathbf{x}_e - \mathbf{X}_n\|} \tag{2.3}$$

The electron charge is represented by q_e and the nuclei charge by $q_n = |q_e| \cdot Z$ (where Z is the atomic number).

The time independent Schrödinger equation (from which one could determine the ground state, perform quantum static computations or accomplish separated representations of the time-dependent solution) is written:

$$-\frac{\hbar^2}{2m_e}\sum_{e=1}^{e=N} \nabla_e^2 \Psi + \sum_{e=1}^{e=N-1}\sum_{e'=e+1}^{e'=N} V_{ee'}\Psi + \sum_{e=1}^{e=N}\sum_{n=1}^{n=M} V_{en}\Psi = E\,\Psi \tag{2.4}$$

where the ground state corresponds to the eigenfunction Ψ_0 associated with the most negative eigenvalue E_0.

Several techniques have been proposed for solving this equation. Some of them lie in the direct solution of the (time-independent or time-dependent) Schrödinger equation. Due to the curse of dimensionality its solution was only possible for very reduced populations of electrons.

Other solution strategies are based on the Hartree-Fock (HF) approach and its derived approaches (post-Hartree-Fock methods). The main assumption of this approach lies in the approximation of the joint electronic wavefunction (related to the N electrons) as a product of N 3D-functions (the molecular orbitals) verifying the antisymmetry restriction derived from the Pauli's principle. Thus, the original HF approach consists of writing the joint wavefunction from a single Slater's determinant. The Schödinger equation allows computing the N molecular orbitals after solving the resulting strongly nonlinear problem. This technique has been extensively used in quantum chemistry to analyze the structure and behavior of molecules involving a moderate number of electrons. Of course, the HF assumption represents sometimes a too crude approximation which invalidates the derived results.

To circumvent this crude approximation different multi-determinant approaches have been proposed. Interested readers can refer to the excellent overview of Cancès et al. [1] as well as the different chapters of the handbook on computational chemistry [2]. The simplest possibility consists in writing the solution as a linear combination of some Slater determinants built by combining n molecular orbitals, with $n > N$. These molecular orbitals are assumed known (e.g. the orbitals related to the hydrogen atom) and the weights are searched to minimize the electronic energy. When the molecular orbitals are built from the Hartree-Fock solution (by employing the ground state and some excited eigenfunctions) the technique is known as the Configuration Interaction method (CI). A more sophisticated technique consists in writing this many-determinants approximation of the solution by using a number of molecular orbitals n (with $n > N$) assumed unknown. Thus, the minimization of the electronic energy leads to compute simultaneously the molecular orbitals as well as the associated coefficients of this many-determinants expansion. Obviously, each one of these unknown molecular orbitals are expressed in an appropriate functional basis (e.g. gaussian functions, ...). This strategy is known as a Multi-Configuration Self-Consistent Field (MCSCF).

All the just mentioned strategies (and others like the coupled cluster or the Moller-Plesset perturbation methods) belong to the family of the wavefunction based methods. In any case all these methods can be only used to solve quantum systems composed of a moderate number of electrons. As we confirm later the main difficulty is not in the dimensionality of the space, but in the use of the Slater determinants (needed to account for the Pauli's principle) whose complexity scales on the factorial of the number of electrons, i.e. in $N!$

The second family of approximation methods, widely used in quantum systems composed of hundreds, thousands and even millions of electrons, are based on the density functional theory (DFT). These models, more than looking for an expression of the wavefunction (with the associated multi-dimensional issue) look for the electronic distribution $\rho(\mathbf{x})$ itself. The main difficulties of this approach are related to the expressions of both the kinetic energy of electrons and the inter-electronic repulsion energy. The second term is usually modelled from the electrostatic self-interaction energy of a charge distribution $\rho(\mathbf{x})$. On the other hand the kinetic energy term is also evaluated in an approximate manner (from the electronic distribution itself in the Thomas-Fermi and related orbital-free DFT models or from a system of N non-interacting electrons—Kohn-Sham models). Obviously, due to the just referred approximations introduced in the kinetic and inter-electronic interaction energies, a correction term is needed, the so-called exchange-correlation-residual-kinetic energy. However, no exact expression of this correction term exists and so different approximate expressions have been proposed and used. Thus, the validity and accuracy of the computed results will depend on the accuracy of the exchange-correlation term that must be fitted for each system.

The models related to the Thomas-Fermi approximation, less accurate in practice because the too phenomenological expression of the kinetic energy coming from the reference system of an uniform non-interacting electron gas, allows us to consider large multi-electronic systems. In a recent work, Gavini et al. [3] performed multi-million atom simulations by employing the Thomas-Fermi-Weizsacker family of orbital-free kinetic energy functionals. On the other hand, the Kohn-Sham based models are a priori more accurate, but they need the computation of the N eigenfunctions related to the N lowest eigenvalues of a non-physical atom composed of N non-interacting electrons.

Transient solutions are very common in the context of quantum gas dynamics (physics of plasma) but are more infrequent in material science when the structure and properties of molecules or crystals are concerned. For this reason, in what follows, we will focus on the solution of the time-independent Schrödinger equation which leads to the solution of the associated multidimensional eigenproblem, whose eigenfunction related to the most negative eigenvalue constitutes the ground state of the system.

Quantum chemistry calculations performed in the Born-Oppenheimer setting consist either (i) in solving the geometry optimization problem, that is, to compute the equilibrium molecular configuration (nuclei distribution) that minimizes the energy of the system, finding the most stable molecular configuration that determines numerous properties like for instance infrared spectrum or elastic constants; or (ii) in performing an *ab initio* molecular dynamics simulation, that is, to simulate the time evolution of the molecular structure according to the Newton law of classical mechanics. Molecular dynamics simulations allow us to compute various transport properties (thermal conductivity, viscosity, ...) as well as some other non-equilibrium properties.

For more details on the mathematical aspects of these models the interested reader can refer to [4–6] and the references therein.

2.1.2 From ab initio to Molecular Dynamics

Depending on the choice of the method, on the accuracy required, and on the computer facility available, the *ab initio* methods allow today for the simulations of systems up to ten, one hundred or some million atoms. In time dependent simulations, they are only convenient for small-time simulations, say not more than a picosecond. However, sometimes larger systems are concerned, and for this purpose one must focus on faster approaches, obviously less accurate. Two possibilities exist: the semi-empirical and the empirical approaches. The semi-empirical approaches speed up the *ab initio* methods by profiting from the information coming from experiments or previous simulations. Empirical methods proceed by considering explicitly only the nuclei, by introducing "empirical" potentials leading to the forces acting on the nuclei. Thus, in the stationary setting only the stable configuration is searched, and for this a geometrical optimization (to computed the nuclei equilibrium distribution) is addressed leading to so-called molecular mechanics. The transient setting results in the classical molecular dynamics but now the computation is sped up by many orders of magnitude with respect to the molecular dynamics where the potentials are computed at the *ab initio level*.

Thus, if we assume a population of M nuclei (of mass m_n) and a two-body potential (many-body potentials are also available), now Newton's law is written for a generic nuclei n:

$$m_n \frac{d^2\mathbf{X}_n}{dt^2} = \sum_{k=1, k \neq n} \mathbf{F}_k^n, \quad \forall n \in [1, \ldots, M] \tag{2.5}$$

where F_k^n denotes the force acting on nucleus n originated by the presence of nucleus k. Obviously these forces can be computed from the gradient of the assumed inter-particles potentials.

Accurate algorithms for integrating these equations exist. The simplectic Verlet's scheme is one of the most used. Molecular dynamics simulations are confronted, despite its conceptual simplicity, with diverse difficulties of different nature:

- The first and most important comes, as previously indicated, from the impossibility of using an "exact" interaction potential derived from quantum mechanics. This situation is particularly delicate when we are dealing with some irregular nuclei distributions such as the ones encountered in the neighborhood of defaults in crystals (dislocations, crack tips, etc.), interfaces between different materials or in zones where different kinds of nuclei coexist.
- The second one comes from the units involved in this kind of simulations: the nuclei displacements are in the nanometric scale, the energies are of the order of the electron-volts, the time steps are of the order of 10^{-15} s. Thus, because of the limits in the computers precision, a change of units is required, which can be easily performed.
- In molecular dynamics the behavior of atoms and molecules is described in the framework of classical mechanics. Thus, the particles energy variations are continuous. The applicability of MD depends on the validity of this fundamental hypothesis. When we consider crystals at low temperature the quantum effects (implying discontinuous energy variations) are preponderant, and in consequence the matter properties at these temperatures cannot be determined by MD simulations. The use of MD is restricted to temperatures higher than the Debye's temperature (for

example the Debye's temperature for the Fe $\alpha \approx 460$ K). This analysis is in contrast to the vast majority of MD simulations carried out nowadays. In fact, higher is the temperature (kinetic energy) and higher results the velocity of particles, requiring shorter time steps in order to ensure the stability of the integration scheme. For this reason, nowadays most of the MD simulations in solid mechanics are carried out at zero degrees Kelvin or at very low temperatures but, as just pointed out, at these temperatures the validity of the computed MD solutions are polluted by the non-negligible quantum effects.

- The prescription of boundary conditions is another delicate task. If the analysis is restricted to systems with free boundary conditions, then the MD simulation can be carried out without any particular treatment. In the other case we must consider a system large enough to ensure that in the analyzed region the impact of the free surfaces can be neglected. Another possibility lies in the prescription of periodic boundary conditions, where an atom leaving the system for example through the right boundary is re-injected in the domain through the left boundary. Moreover, the particles located in the neighborhood of a boundary are influenced by the ones located in the neighborhood of the opposite boundary. The imposition of other boundary conditions is more delicate from both the numerical and the conceptual points of view. For example, what is the meaning of prescribing a displacement on a boundary? Each situation requires a deep analysis in order to define the best (the most physical) way to prescribe the boundary conditions.
- There are other difficulties related to the transient analysis. We consider a thermal system in equilibrium, i.e. a system in which the velocities follow the Maxwell-Boltzmann distribution. Now, we proceed to heat the system. One possibility lies in increasing suddenly the kinetic energy of each particle. Obviously, even if the resulting velocities define a Maxwell-Boltzmann's distribution, the system remains off equilibrium because the partition between kinetic and potential energies is not the appropriate one. For this reason we must proceed to relax the system that evolves from this initial (non-physical) state to the equilibrium one. Another (more physical) possibility lies in the incorporation of a large enough ambient region around the analyzed system, whose particles are initially in equilibrium at the highest temperature. Now, both regions (the system and the ambient) interact, and the system initiates its heating process that reaches its equilibrium some time later. The final state of both evolutions is the same, but the time required to reach it depends on the technique used to induce the heating. The first transient is purely numerical whereas the second one is more physical allowing the identification of some transport coefficients (for example the thermal conductivity).
- Finally the CPU time continues to be the main limitation of MD simulations. The strongest handicap is related to the necessity of considering at each time step and for each particle the influence of all the others particles. Thus, the integration method seems to scale with the square of the number of particles. Even if some computing time savings can be introduced in the neighbor search, the extremely small time steps and the extremely large number of particles required to describe real scenarios, limit considerably the range of applicability of this kind of simulations, that has been accepted to be nowadays of the order of a cubic micrometer,

even when the systems are considered as very low, and then non-physical, temperatures (close to zero degrees Kelvin). We can notice that, despite the impressive advances in the computational availabilities, the high performance computing and the use of massive parallel computing platforms, the state of the art does not allow the treatment of macroscopic systems encountered in practical applications of physics, chemistry and engineering.

The above mentioned difficulties to perform fully molecular dynamics (MD) simulations motivated the proposal of hybrid techniques that apply MD in the regions where the unknown field varies in a non-uniform way (molecular dynamics model) and a standard finite element approximation in those regions where the unknown field variation can be considered as uniform (continuous model). The main questions concerned by these bridging strategies concern: (i) the kinematics representations in both models; (ii) the transfer conditions on the MD and continuous models interface and (iii) the macroscopic constitutive equation employed in the continuous model.

Different alternatives exist, and the construction of such bridges is nowadays one of the most active topics in computational mechanics. The spurious reflection of the high frequency parts of the waves is one of the main issues. We would like only to mention three "families" of bridging techniques, giving some key references in which the interested reader could find other extremely valuable references: (i) the quasi-continuum method proposed by Tadmor and Ortiz [7]; (ii) the multi-scale method proposed by Wagner and Liu [8] and (iii) the methods based on the "Arlequin" approach [9] like the one proposed by Belytschko in [10].

2.1.3 Coarse Grained Modeling: Brownian Dynamics

Sometimes one is interested in analyzing the behavior of a system composed by a series of microscopic entities (particles assumed with a null extension) dispersed into another fluid (the solvent). The kinematics of such particles depends, of course, on their interactions with the solvent particles. A real molecular dynamics simulation is definitively forbidden (the nowadays MD simulation feasibilities rarely exceeds the number of particles contained within a cube of one micron of side).

One possibility to reduce the size of the discrete models lies in considering only the particles of interest. The other particles (the ones that constitute the solvent) are not considered explicitly and only their averaged effects are retained in the modeling.

Thus, the motion equation of a particle whose position is described by $\mathbf{x}_i$, is governed by the Langevin equation:

$$m\frac{d^2\mathbf{x}_i}{dt^2} = \xi\left(\frac{d\mathbf{x}_i}{dt} - \mathbf{v}_f(\mathbf{x}_i)\right) + \mathbf{F}_i^{ext}(t); \quad \forall i, \tag{2.6}$$

where m denotes the particle mass, $\mathbf{x}_i$ the position of particle i, ξ the friction coefficient, $\mathbf{v}_f(\mathbf{x}_i)$ the fluid velocity at position $\mathbf{x}_i$ and $\mathbf{F}_i^{ext}(t)$ all the other forces acting on the particle i (coming from a external potential or from the solvent particles bombardment). We can notice that even if this model doesn't incorporate explicitly the solvent particles population, their effects are taken into account from the drift term $\xi\left(\frac{d\mathbf{x}_i}{dt} - \mathbf{v}_f(\mathbf{x}_i)\right)$ as well as by the impact forces.

In the last expression the drift term is quite standard, however the external forces acting on each particle deserve some additional comment. In what follows and without any detriment of generality we assume that there are no other forces than the one coming from the solvent particles bombardment and that the solvent is macroscopically at rest, i.e., $\mathbf{v}_f = \mathbf{0}$. The random nature of the interaction force is modelled from a statistical distribution function that becomes fully defined as soon as its mean value and its standard deviation are fixed. Concerning the mean, one expects a null value if the microscopic dynamics are isotropic. Concerning the standard deviation one must proceed within the statistical mechanics framework. In what follows we summarize the main ideas for the derivation of the standard deviation expression.

Let us define $B_{\Delta t}$

$$B_{\Delta t} = \int_0^{\Delta t} \frac{F^{ext}(t)}{m} dt = \sum_{k=1}^{k=p} \frac{F^{ext}(t_k)}{m} \delta t, \tag{2.7}$$

where we assume that in Δt the particle is subjected to p impacts from the solvent particles, with $p \gg 1$. These impacts are modeled by a constant force $F^{ext}(t_k)$ that applies for a time δt ($\Delta t = p\delta t$). By invoking the central limit theorem we conclude that $B_{\Delta t}$ follows a gaussian distribution $\mathcal{N}(0, q\Delta t)$ because the number of impacts scales linearly with Δt.

Now, to compute q, one could integrate the Langevin's equation to obtain the equilibrium velocity distribution W:

$$W\left(\frac{dx}{dt}, t \to \infty\right) = \sqrt{\frac{\xi}{m\pi q}}\, e^{\frac{\xi\left(\frac{dx}{dt}\right)^2}{mq}} \tag{2.8}$$

that must coincide with the Maxwell-Boltzmann one (canonical ensemble), from which we deduce an expression of q:

$$q = \frac{2\xi k_b T}{m^2} \tag{2.9}$$

where k_b is the Boltzmann constant and T the absolute temperature.

Thus, the Langevin equation is fully defined, by writing:

$$B_{\Delta t} = \int_0^{\Delta t} \frac{F^{ext}(t)}{m} dt = \mathcal{N}\left(0, 2\frac{\xi k_b T}{m^2}\Delta t\right) \tag{2.10}$$

an expression that applies also in the presence of other forces like the ones coming from a gradient of a potential or when $\mathbf{v}_f \neq \mathbf{0}$.

Thus, one could track the movement of each particle $\mathbf{x}_i$ by considering a standard drift term and a random force whose distribution is perfectly known. This stochastic approach has been traditionally also used for solving deterministic models as the advection-diffusion one, because one could compute some moments of the resulting distribution by tracking a moderate population of particles instead of discretizing the deterministic counterpart of the advection-diffusion model by using one of the standard mesh-based discretization techniques that could involve in the case of 3D models an excessive number of degrees of freedom.

To illustrate this procedure we consider the simplest form of the advection-diffusion equation:

$$\frac{\partial C}{\partial t} + \mathbf{v} \cdot \nabla C = D \Delta C \tag{2.11}$$

where $C = C(\mathbf{x}, t)$ is the concentration field and D is the so-called diffusion coefficient.

Obviously, this simple parabolic equation could be solved by using any standard technique (finite differences, finite elements, spectral methods, finite volumes, the method of particles, ...), but in what follows we are solving it using a stochastic approach. For this purpose we assume the initial condition represented by N-Dirac masses:

$$C(\mathbf{x}, t = 0) \approx C^0(\mathbf{x}) = \sum_{i=1}^{i=N} c_i \, \delta(\mathbf{x} - \mathbf{x}_i(t = 0)) \tag{2.12}$$

Here, we are not discussing the choice of the coefficients c_i and the locations $\mathbf{x}_i$. Several possibilities exist to perform a choice trying to represent, as precisely as possible, the initial concentration distribution. Most of them proceed by regularizing the Dirac distribution and then enforcing the minimization of $\| C(\mathbf{x}, t = 0) - C^0(\mathbf{x}) \|$. From now on, c_i and $\mathbf{x}_i(t = 0)$ are assumed known. Moreover, as the considered advection-diffusion equation does not contain source terms, the weights c_i remain unchanged during the evolution of the pseudo-particles positions $\mathbf{x}_i(t)$.

Now, the simplest explicit integration algorithm proceeds by updating the particle position considering both the deterministic and the random contributions:

$$\mathbf{x}_i^{n+1} = \mathbf{x}_i^n + \mathbf{v}(\mathbf{x}_i^n)\Delta t + \mathcal{N}(0, 2D\Delta t)\mathbf{u}, \quad \forall i \tag{2.13}$$

where $\mathbf{x}_i^{n+1} \equiv \mathbf{x}_i(t = (n+1) \cdot \Delta t)$ and $\mathbf{u}$ is a unit random vector.

Now the distribution moments can be easily computed, and the concentration field could be reconstructed by employing some appropriate smoothing.

It is usual to find this kind of advection-diffusion equations in many branches of science and engineering. In particular they are encountered when one models macromolecular materials within the statistical mechanics framework, as we describe later. Despite its intrinsic simplicity, sometimes they arise defined in highly dimensional spaces including the physical and the conformation coordinates. To avoid the curse of dimensionality drawback characteristic of mesh-based techniques, different authors proposed the use of stochastic techniques exploiting the equivalence between the so-called Fokker-Planck equation and the Ito's stochastic equation [11], similar to the just described equivalence between models (2.11) and (2.13). Thus, if one is only interested in computing some moments of the resulting distribution function a moderate population of particles is enough to describe accurately the evolution of such moments. The size of the pseudo-particles population that must be considered for computing accurately the different moments of the solution scales linearly with the dimension of the space, however, if one wants to reconstruct the distribution itself, an impressive number of particles is required which do not scale anymore linearly with the dimension of the space.

In general the technique just presented, is conceptually very simple and then easy to implement in a computer or in a parallel computing platform. Explicit integration schemes are usually employed, needing for a careful choice and control of the time step.

It is nowadays widely recognized that the main drawbacks of stochastic techniques are: (i) the control of the statistical noise that makes difficult the use of the stochastic approach to perform inverse parameter identification or optimization, because the poor accuracy in the sensitivity analysis; (ii) the difficulty to reconstruct the model solution itself even in moderate multidimensional models; and (iii) the necessity to solve always the transient model, even if one is only interested in the steady state.

Moreover, in complex flows simulation using for example a finite element solver for the flow kinematics computation, one must ensure that all the elements contains a number of particles, at least enough to allow computing the virial stress (the usual micro-macro bridge). Different possibilities exist, but all of them have a non-negligible impact on the solution. Thus, if new particles are added (and probably others removed) the size of the model is changing, tracking procedures are time consuming, and the initialization of the just introduced particles induces a noticeable numerical diffusion.

One could think that the aforementioned difficulties could be circumvented by employing a Lagrangian description of the flow combined with a Lagrangian description of the microstructure evolution (stochastic approach), however usual mesh-based strategies fail because the high distortion of the meshes during the flow. A first tentative strategy of coupling a meshless Lagrangian description of the flow kinematics (using the natural element method widely described in [12, 13]) and a Lagrangian microstructure description have been recently performed in [14]. This technique could be extended for coupling a Lagrangian flow description with a stochastic description of the microstructure evolution. In any case the difficulties just mentioned will persist.

To reduce the computational cost of numerical simulations different model reduction techniques have been recently proposed [15]. However, the coupling of such techniques (based on the proper orthogonal decomposition—also known as Karhunen-Loève decomposition) with a Lagrangian description of the microstructure evolution remains nowadays an open problem. However, sometimes the stochastic model can be written in an Eulerian form (Brownian Configuration Fields) and in that case, as illustrated in [16], the model reduction runs opening some interesting perspectives.

2.1.4 Coming Back to Continuous Descriptions: Kinetic Theory Models

The next level of description concerns statistical mechanics where the knowledge of individuals is substituted by a kind of averaged knowledge described by a probability distribution function that depends on some conformational coordinates depending on the considered model. In what follows we address some models involving more or less large conformational spaces.

2.1.4.1 The Vlasov-Poisson-Boltzmann Equation

Firstly, we consider the dynamics of N electrically charged particles of mass m. When N becomes too large, direct molecular dynamics simulations result to be prohibitive from the computing time viewpoint. Thus, more than describing the system from the position and velocities of all the particles, one could introduce the function $f(t, \mathbf{x}, \mathbf{v})$ given the number of particles that, at time t, are located within an elementary volume $d\mathbf{x} = (dx, dy, dz)^T$ placed at position $\mathbf{x}$ and having velocities within the volume defined by $d\mathbf{v} = (du, dv, dw)^T$ around $\mathbf{v}$. Now, the density balance is written:

$$\frac{\partial f}{\partial t} + \mathbf{v} \cdot \nabla_x f + \mathbf{a}(\mathbf{x}, t) \cdot \nabla_v f = S(t, \mathbf{x}, \mathbf{v}), \tag{2.14}$$

where ∇_x and ∇_v represent the gradient operator in the physical and velocity spaces respectively. We assume that the acceleration $\mathbf{a} = \frac{d\mathbf{v}}{dt}$ does not depend on the velocity (for this reason it is is not affected by the velocity-gradient ∇_v. The source term $S(t, \mathbf{x}, \mathbf{v})$ represents the so-called collision term and can be derived from an appropriate physical analysis.

We do not need any physics to model the velocity field $\mathbf{v}$ because now the velocity field is a real coordinate, like the spatial ones. On the contrary, we need to define the acceleration field $\mathbf{a}(\mathbf{x}, t)$. For this purpose we consider the Newton's law $\mathbf{a} = \frac{\mathbf{F}}{m}$, and compute the force acting on the particles by taking into account the nature of the system, that in the case here addressed consists of a population of charged particles interacting by means of the Coulomb's potential. The electrostatic potential $U(\mathbf{x}, t)$ can be computed by solving the associated Poisson's problem:

$$\Delta U(\mathbf{x}, t) = -4\pi k Q(\mathbf{x}, t), \tag{2.15}$$

where k is the Coulomb law constant and $Q(\mathbf{x}, t)$ is the electrical charge inside the volume $d\mathbf{x}$ around $\mathbf{x}$ that can be computed from:

$$Q(\mathbf{x},t) = q \int_{\mathbb{R}^3} f(\mathbf{x},\mathbf{v},t)d\mathbf{v}, \tag{2.16}$$

with q the particle charge. The force acting on a particle can be finally computed by using:

$$\mathbf{F}(\mathbf{x},t) = -q\ \nabla_x U(\mathbf{x},t) \tag{2.17}$$

Remark 2.1.1

1. When the particles are not charged the steady state solution of Eq. (2.14) leads to the Maxwell-Boltzmann distribution when the appropriate choice of the collision term representing the particles collisions is made.
2. When the interaction potential does not comply the Coulomb's law, model (2.15)–(2.17) cannot be employed. In this case we must compute the force by using:

$$\mathbf{F}(\mathbf{x},t) = \int_{\mathbb{R}^3} \mathbf{F}(\mathbf{x},t;\mathbf{x}')\ \tilde{f}(\mathbf{x}',t)d\mathbf{x}' \tag{2.18}$$

   where

$$\tilde{f}(\mathbf{x},t) = \int_{\mathbb{R}^3} f(\mathbf{x},\mathbf{v},t)d\mathbf{v} \tag{2.19}$$

   and $\mathbf{F}(\mathbf{x},t;\mathbf{x}')$ represents the force at position $\mathbf{x}$ originated by a particle located at position $\mathbf{x}'$. In any case this analysis fails when (i) the inter-particle potential leads to a non-definite integral (2.18); and (ii) in the case of dense systems where the movements of particles is correlated.
3. When the acceleration does not depend on the velocity (as was assumed in Eq. (2.14)) and the collision terms vanishes, an equivalence between the conservation equation and the Liouville's theorem in the phase space can be established.
4. In general the establishment of collision terms is quite difficult. To circumvent this difficulty and assuming that the equilibrium distribution is known $f_{eq}(\mathbf{x},\mathbf{v},t)$, one could approximate the collision term by

$$S(t,\mathbf{x},\mathbf{v}) = \frac{f_{eq}(\mathbf{x},\mathbf{v},t) - f(\mathbf{x},\mathbf{v},t)}{\tau} \tag{2.20}$$

   where τ represents a relaxation time. This approximation leads to the so-called BFK models.
5. Equation (2.14), also known as the Vlasov-Poisson-Boltzmann equation, is widely used to model quantum plasmas. One could imagine the extension of this formalism (within the BBGKY hierarchy) to a variety of physical models: colloids, ferrofluids, coarse grained molecular dynamics, crystallization, demixing, etc. The associated kinetic theory descriptions constitute the Vlasov-Fokker-Planck models.

6. The kinetic theory formalism allows transforming a discrete model into its continuous counterpart. However, in general, the continuous descriptions involve highly multidimensional spaces and their evolutions are governed by hyperbolic non-linear partial differential equations. To solve these kind of models appropriate stabilized solvers, able to proceed in highly multidimensional spaces, are needed.

In the previous paragraphs we introduced some ideas related to kinetic theory models of systems composed of particles. We introduce in the next paragraphs some models describing the microscopic modeling of polymeric liquids.

2.1.4.2 Kinetic Theory Description of Polymeric Liquids.

For the sake of simplicity we focus on polymer solutions (the entangled systems related to the polymer melts can be also described in the kinetic theory framework [17, 18]). We address the Bead-Spring-Chain (BSC) model of polymer solutions. The BSC chain consists of $S+1$ beads connected by S springs. The bead serves as an interaction point with the solvent and the spring contains the local stiffness information depending on local stretching (see [17] for more details). From now on we are also assuming a fully developed homogeneous flow. Thus, the microstructural state does not depend on the space coordinates.

The dynamics of the chain is governed by viscous drag, Brownian and connector forces. If we denote by $\dot{\mathbf{r}}_k$ the velocity of bead k and by $\dot{\mathbf{q}}_k$ the velocity of the spring connector $\mathbf{q}_k$, we have

$$\dot{\mathbf{q}}_k = \dot{\mathbf{r}}_{k+1} - \dot{\mathbf{r}}_k \qquad \forall k = 1, \ldots, S. \tag{2.21}$$

The dynamics of each bead can be written as:

$$\underbrace{-\zeta(\dot{\mathbf{r}}_k - \mathbf{v}_0 - \nabla\mathbf{v}\cdot\mathbf{r}_k)}_{Viscous\ drag} - \underbrace{k_b T\frac{\partial ln(\psi)}{\mathbf{r}_k}}_{Brownian\ effects} + \underbrace{\mathbf{F}_k^c - \mathbf{F}_{k-1}^c}_{Connector\ forces} = 0, \tag{2.22}$$

where ζ is the drag coefficient, $\mathbf{v}$ is the velocity field, $\mathbf{v}_0$ is an average velocity, k_b is the Boltzmann constant, T is the absolute temperature and ψ is the distribution function $\psi(\mathbf{r}_1, \ldots, \mathbf{r}_{S+1}, t)$. From Eqs. (2.21) and (2.22) we obtain:

$$\dot{\mathbf{q}}_k = \nabla\mathbf{v}\cdot\mathbf{q}_k - \frac{1}{\zeta}\sum_{l=1}^{S}\mathbf{A}_{kl}\cdot\left(k_b T\frac{\partial ln(\psi)}{\partial\mathbf{q}_l} + \mathbf{F}_l^c\right), \tag{2.23}$$

where $\mathbf{A}_{kl}$ are the components of the Rouse matrix (see [17] for more details).

In the Rouse model the connector force $\mathbf{F}^c$ is a linear function of the connector vector, but we could use Finitely Extensible Nonlinear Elastic (FENE) springs, with a dimensionless connector force given by:

$$\mathbf{F}^c(\mathbf{q}_k) = \frac{1}{1 - \mathbf{q}_k^2/b}\mathbf{q}_k, \tag{2.24}$$

where $\sqrt{b}$ is the maximum dimensionless length of each spring connector of the chain.

Introducing Eq. (2.23) in the equation governing the evolution of the distribution function, and considering homogeneous flows,

$$\frac{\partial \psi(\mathbf{q}_1, \ldots, \mathbf{q}_S, t)}{\partial t} = -\sum_{k=1}^{S} \left(\frac{\partial}{\partial \mathbf{q}_k} (\dot{\mathbf{q}}_k \psi(\mathbf{q}_1, \ldots, \mathbf{q}_S, t)) \right), \tag{2.25}$$

we obtain

$$\frac{\partial \psi}{\partial t} = -\sum_{k=1}^{S} \left(\frac{\partial}{\partial \mathbf{q}_k} \left((\nabla \mathbf{v} \cdot \mathbf{q}_k - \frac{1}{\zeta} \sum_{l=1}^{S} \mathbf{A}_{kl} \cdot \mathbf{F}_l^c)\, \psi \right) \right) + \frac{k_b T}{\zeta} \sum_{k=1}^{S} \sum_{l=1}^{S} \mathbf{A}_{kl} \frac{\partial^2 \psi}{\partial \mathbf{q}_k \partial \mathbf{q}_l}. \tag{2.26}$$

The micro–macro bridging is performed by computing the virial stress, that within the rheology community is known as Kramer's rule. The main difficulty in using this description is the highly multidimensional problem defined by Eq. (2.26) that needs specific advanced solvers such as the ones that we proposed in some of our former works within the PGD framework.

2.1.4.3 Kinetic Theory Description of Rod-Like Suspensions in Complex Flows

In the case of a dilute suspension of rod-like particles (short fibers, nanofibers, functionalized carbon nanotubes or even rod-like molecules), the configuration distribution function (also known as orientation distribution function) gives the probability of finding the particle in a given direction. Obviously, this function depends on the physical coordinates (space and time) as well as on the configuration coordinates, that taking into account the rigid character of the particles, are defined on the surface of the unit sphere. Thus, we can write $\psi(\mathbf{x}, t, \mathbf{p})$, where $\mathbf{x}$ defines the position of the rod center of mass, t the time and $\mathbf{p}$ the unit vector defining the rod orientation. The evolution of the distribution function is given by the Fokker-Planck equation

$$\frac{d\psi}{dt} = -\frac{\partial}{\partial \mathbf{p}}(\psi \dot{\mathbf{p}}) + \frac{\partial}{\partial \mathbf{p}}\left(D_r \frac{\partial \psi}{\partial \mathbf{p}}\right) \tag{2.27}$$

where d/dt represents the material derivative, D_r is a diffusion coefficient and $\dot{\mathbf{p}}$ is the particle rotation velocity. The orientation distribution function must verify the normality condition:

$$\int_{S(0,1)} \psi(\mathbf{p}) d\mathbf{p} = 1 \tag{2.28}$$

where $S(0, 1)$ denotes the surface of the unit sphere.

For ellipsoidal particles and when the suspension is dilute enough, the rotation velocity can be obtained from the Jeffery's equation

$$\dot{\mathbf{p}} = \boldsymbol{\Omega} \cdot \mathbf{p} + k\, \mathbf{D} \cdot \mathbf{p} - k(\mathbf{p}^T \cdot \mathbf{D} \cdot \mathbf{p})\, \mathbf{p} \tag{2.29}$$

where $\boldsymbol{\Omega}$ and $\mathbf{D}$ are the vorticity and the strain rate tensors respectively, associated with the fluid flow undisturbed by the presence of the suspended particles, and k is a scalar which depends on the particle aspect ratio λ (ratio of its length and diameter)

$$k = \frac{\lambda^2 - 1}{\lambda^2 + 1}, \tag{2.30}$$

that for rod-like particles can be assumed $k \approx 1$.

In complex flows simulations the solution of the Fokker-Planck equation involves some numerical difficulties related to: (i) its multidimensional character, i.e., $\psi(\mathbf{x}, t, \mathbf{p}) : \Omega \subset \mathbb{R}^3 \times \mathbb{R}^+ \times S(0, 1) \to \mathbb{R}^+$; (ii) the geometrical complexity of the physical domain Ω ; (iii) its purely advective character in the physical space; and (iv) the advection effects in the conformation space. Both last behaviors need appropriate numerical stabilizations (e.g. upwinding) of discrete models.

2.2 Advanced Solvers for Multi-dimensional Models

The solution of models like the one just addressed in Eq. (2.26) needs new advanced numerical strategies, because the standard ones suffer the curse of dimensionality. Some strategies have been recently proposed for solving models defined in multi-dimensional spaces. The sparse grid techniques are one of the most popular [19], but as we described in the first section, separated representations were also used for solving the models encountered in quantum mechanics (Hartree-Fock and post-Hartree-Fock techniques).

The sparse grids [19] are restricted (as argued in [20]) for treating models involving up to twenty dimensions. On the other hand, separated representations like the one considered in the multi-configuration-self-consistent-fields [2]. We retained the PGD technology thoroughly described in the previous chapter that allows circumventing efficiently the curse of dimensionality.

2.3 Numerical Examples

In this section we illustrate the application of separated representations to the solution of different models defined in highly multidimensional spaces suffering the so-called curse of dimensionality encountered in the finest description of materials.

2.3.1 On the Separated Representation of the Schrödinger Equation

Finally we consider a challenging problem, the one associated with the solution of the Schrödinger equation introduced in Sect. 2.1.1.

In this case to circumvent the curse of dimensionality that this model involves we could try to perform, as in the previous examples, a separated representation of the wavefunction Ψ.

In what follows we consider a system composed of N electrons and a single nucleus with atomic number $Z = N$. In this case one could assume the time-independent wavefunction approximation:

$$\Psi(\mathbf{x}_1, \ldots, \mathbf{x}_N; \mathbf{X}) \approx \sum_{j=1}^{j=n} \alpha_j \, \phi_j^1(\mathbf{x}_1) \cdot \phi_j^2(\mathbf{x}_2) \cdot \cdots \cdot \phi_j^N(\mathbf{x}_N) \tag{2.31}$$

where $\mathbf{X}$ denotes the nucleus position.

However, the Pauli's exclusion principle that applies for fermions (electrons belong to the fermions family) implies the antisymmetry of the wavefunction. To ensure the antisymmetry one could proceed by introducing the Slater's determinants:

$$D_j = \begin{vmatrix} \phi_j^1(\mathbf{x}_1) & \phi_j^2(\mathbf{x}_1) & \cdots & \phi_j^N(\mathbf{x}_1) \\ \phi_j^1(\mathbf{x}_2) & \phi_j^2(\mathbf{x}_2) & \cdots & \phi_j^N(\mathbf{x}_2) \\ \vdots & \vdots & \ddots & \vdots \\ \phi_j^1(\mathbf{x}_N) & \phi_j^2(\mathbf{x}_N) & \cdots & \phi_j^N(\mathbf{x}_N) \end{vmatrix} \tag{2.32}$$

that, introduced in the separated representation

$$\Psi(\mathbf{x}_1, \ldots, \mathbf{x}_N; \mathbf{X}) \approx \sum_{j=1}^{j=n} \alpha_j \, D_j \tag{2.33}$$

ensures the antisymmetry of the resulting wavefunction.

However, the last representation hides two difficulties. The first one is related to the multidimensional character of the model (we must recall that the wavefunction is defined in a space of dimension $3 \times N$) and the second one is related to the complexity in evaluating the Slater determinants that scales in $N!$. In order to isolate each difficulty, we are considering an "imaginary world" in which the electrons are not subjected to the Pauli's exclusion principle. In this case the separated representation (2.31) works and it could be used to solve the Schrödinger equation.

For the sake of simplicity, in what follows, we assume that each electron "lives" in a 1D dimensional space, i.e., $\Psi(x_1, \ldots, x_N; X)$, where $x_k \in \mathbb{R}, \ \forall k$.

Now, we consider different systems containing the more and more electrons: $N = 1, 2, 3, 4, 5, 10, 20, 50$. In the last case the Schrödinger equation is defined in a space of dimension 50.

For each system, the time independent Schrödinger equation (2.4) is solved using the separated representation (2.31), and the ground state eigenvalue Ψ_0 and its associated energy E_0 is computed. Now, the ground state electronic distribution $\rho^0(x)$ can be obtained by calculating the ground state distribution of each electron e, ρ_e^0:

$$\rho_e^0(x) = \int_{\mathbb{R}^{N-1}} |\Psi_0(x_1, \ldots, x_{e-1}, x_{e+1}, \ldots x_N)|^2 \, dx_1 \ldots dx_{e-1} dx_{e+1} \cdots dx_N \tag{2.34}$$

and adding all the electrons' contributions

$$\rho^0(x) = \sum_{e=1}^{e=N} \rho_e^0. \tag{2.35}$$

Figure 2.1 depicts the ground-state electronic distribution for $N = 1$, $N = 2$, $N = 3$, $N = 5$ and $N = 10$. Obviously the electronic distribution is concentrated around the nucleus position and increases as the number of electrons increases. In that figure, to ensure a good quality in the curves resolution we do not include the electronic distributions of the systems composed of 20 and 50 electrons.

It is easy to verify that because the Pauli's exclusion principle does not apply, $\rho^0(N) = N\rho^0(1)$, where $\rho^0(N)$ and $\rho^0(1)$ denote the ground state densities of systems composed of N and 1 electrons respectively. Thus, we can conclude that all the electrons are occupying the same orbital, a kind of s-orbital (that in 3D systems would be spherical).

Figure 2.2 depicts for these "imaginary" systems the evolution of the ground state energy (the most negative eigenvalue related to the solution of Eq. (2.4)) as a function of the number of electrons involved in the system.

From this analysis we conclude that separated representation allows us to circumvent the curse of dimensionality related to quantum mechanics systems, making possible the treatment of multidimensional models. However, the previous analysis concerns "non-real" systems, because electrons are fermions, and for fermions the Pauli's principle applies.

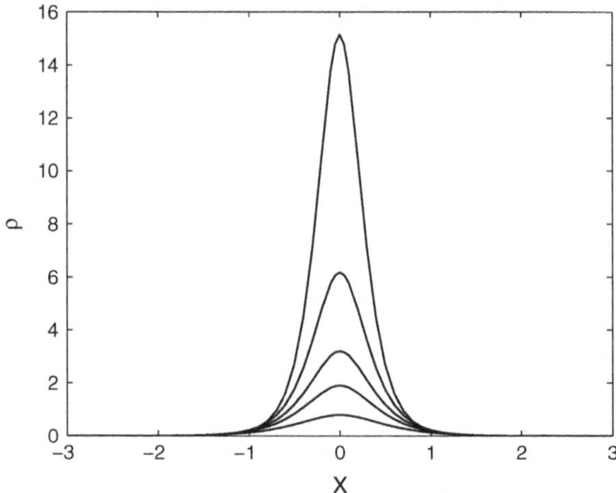

Fig. 2.1 Ground state electronic density for systems composed of 1, 2, 3, 5 and 10 electrons—the Pauli's principle has not been applied

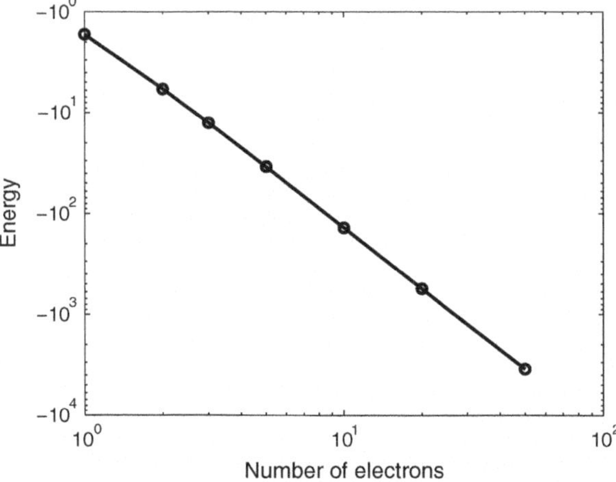

Fig. 2.2 Evolution of the ground state energies with the number of electrons involved in the system—the Pauli's principle has not been applied

Even if separated representations could be applied for solving bosonic quantum systems (the Pauli's principle does not apply for bosons) in material science such systems have a limited interest. For this reason we must come back to the "real world" of electronic systems.

In the present case, the separated representation to be considered is the one making use of the Slater determinants (2.33). All the practical details concerning the discretization of the Schrödinger equation by using a wavefunction separated representation within the Slater's determinants formalism can be found in [21]. In this case, as the complexity scales with the factorial of the number of electrons involved in the quantum system, at present, in our knowledge, only small systems can be solved (containing up to ten electrons).

Figure 2.3 depicts the ground state electronic distribution of systems composed of 1 to 5 electrons and a single nucleus with atomic number $Z = 3$. It also shows the differences between each two successive configurations. We can notice that the first two electrons are located in an s-type orbital (both electrons have different spin which makes it possible). For more than two electrons we appreciate the appearance of p-type orbitals. These p-type orbitals can be identified by subtracting the electronic density functions related to both kind of populations (up to two electrons and more than two electrons).

Figure 2.4 compares the evolution of the ground state energy with the number of electrons N involved in the quantum system composed on a single nucleus with $Z = N$, when the Pauli's principle is or is not taken into account. For comparison purposes, when the Pauli's exclusion principle was activated, the spin coordinate was removed, because in this way each electron is forced to occupy a different orbital (as was the case when the Pauli's principle was not taken into account). We can notice that its consideration increases the value of the ground state energy.

From the previous analysis we can affirm that the remaining unsolved difficulty related to the solution of the Schrödinger equation lies in the antisymmetry restriction that the Pauli's principle addresses for fermions, and that as previously argued, its complexity scales as $N!$, which becomes extremely large even for systems composed of a moderate number of electrons.

2.3.2 *Kinetic Theory Description of Rod-Like Aggregating Suspensions*

In this section we come back to the suspension model summarized in Sect. 2.1.4. In order to increase the model complexity we are assuming that the rods can flocculate creating large aggregates that due to the shear induced by the flow, are continuously broken. Thus, aggregation and disaggregation mechanisms coexist and two populations can be identified: the one related to free rods (pendant population) and the one associated with the aggregated rods (active population). Figure 2.5 depicts both populations and the flow induced aggregation/disaggregation.

The kinetic theory description of such systems contains two coupled Fokker-Planck equations given the orientation distribution of rods belonging to each one of these populations, $\Psi(\mathbf{x}, t, \mathbf{p})$ and $\Phi(\mathbf{x}, t, \mathbf{p})$ for the active and pendant respectively:

$$\frac{d\Psi}{dt} = -\frac{\partial}{\partial \mathbf{p}}(\dot{\mathbf{p}}\Psi) + D_{r1}\frac{\partial^2 \Psi}{\partial \mathbf{p}^2} - V_d\Psi + V_c\Phi, \tag{2.36}$$

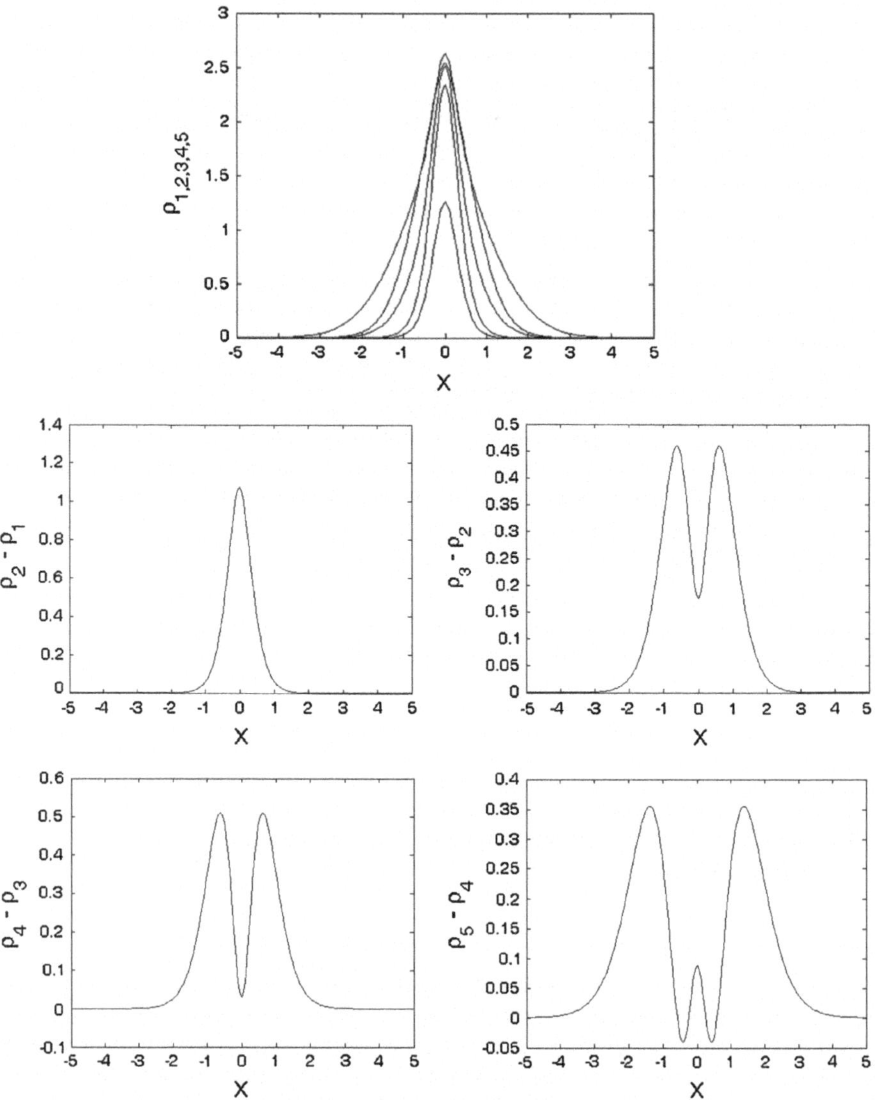

Fig. 2.3 Ground state electronic density for systems composed of 1, 2, 3, 4 and 5 electrons—Pauli principle applies

$$\frac{d\Phi}{dt} = -\frac{\partial}{\partial \mathbf{p}} (\dot{\mathbf{p}}\Phi) + D_{r2}\frac{\partial^2 \Phi}{\partial \mathbf{p}^2} + V_d \Psi - V_c \Phi, \tag{2.37}$$

where V_d and V_c represent the velocity of destruction and construction of the active population respectively.

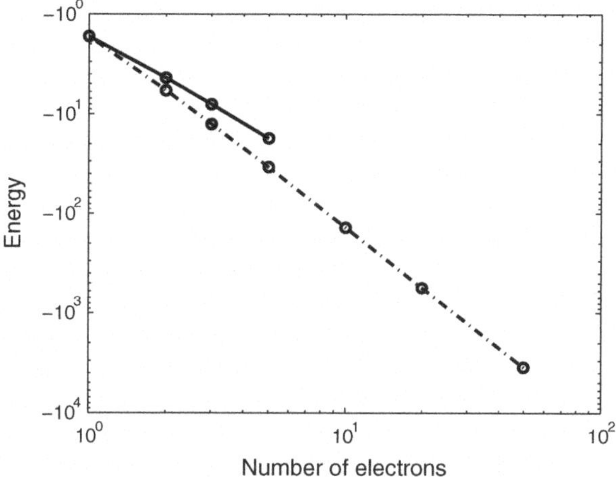

Fig. 2.4 Evolution of the ground state energy when the Pauli's principle is activated

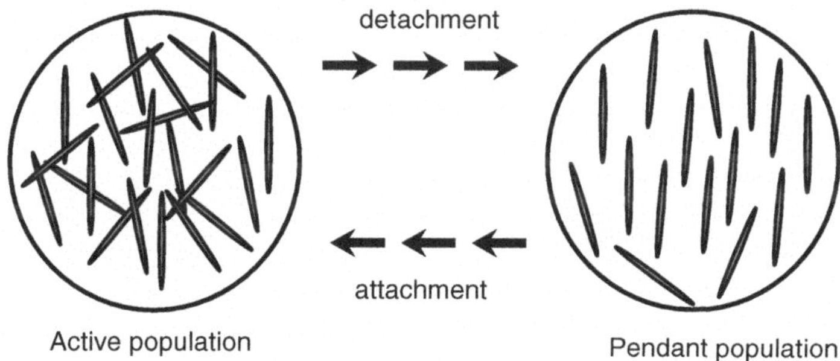

Fig. 2.5 Flow induced aggregation/disaggregation

The normality condition is written

$$\int_{S(0,1)} \left(\Psi\left(\mathbf{x},t,\mathbf{p}\right)+\Phi\left(\mathbf{x},t,\mathbf{p}\right)\right)\,d\mathbf{p}=1, \quad \forall\mathbf{x},\forall t. \tag{2.38}$$

We consider that the flow takes place in a converging channel. The steady state flow kinematics (assumed undisturbed by the presence of the suspended particles) was computed by solving the Stokes equations.

As we are interested in computing the steady state solution, and because the advection character of the Fokker-Planck equations in the spatial coordinates ($\mathbf{x}$), we decided to integrate both coupled Fokker-Planck equations along some particular flow streamlines. The separated representation of both orientation distribution functions writes:

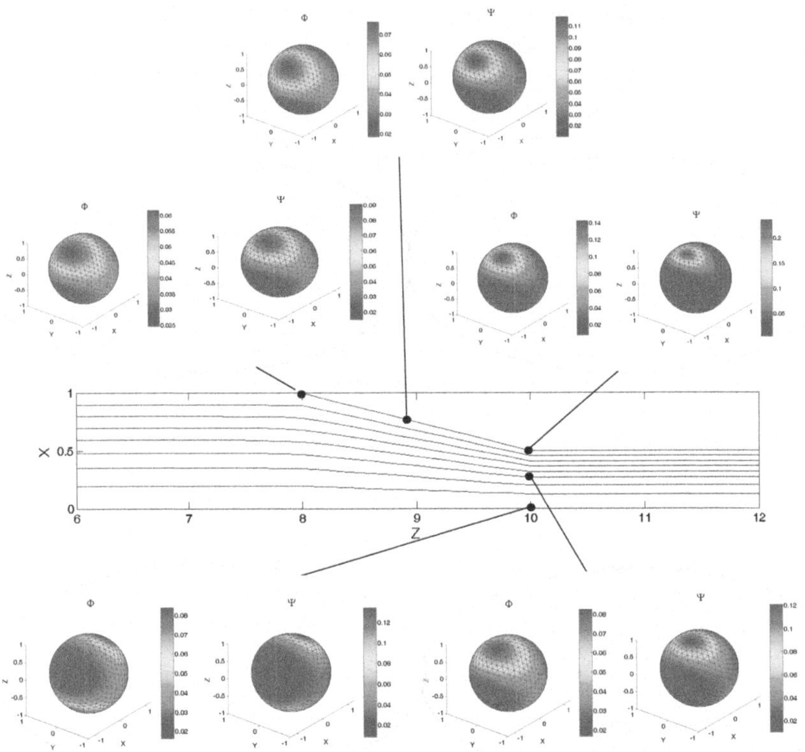

Fig. 2.6 Orientation distribution of active and pendant populations in a contraction flow

$$\begin{pmatrix} \Psi_{st}(s, \mathbf{p}) \\ \Phi_{st}(s, \mathbf{p}) \end{pmatrix} = \sum_{j=1}^{n} \begin{pmatrix} \alpha_j^{st} E_j^{st}(\mathbf{p}) F_j^{st}(s) \\ \beta_j^{st} G_j^{st}(\mathbf{p}) H_j^{st}(s) \end{pmatrix} \tag{2.39}$$

where s denotes thee curvilinear coordinate defining the streamlines, and the index "st" refers to the particular streamline along which the integration is performed.

Figure 2.6 depicts the resulting orientation distributions of both populations at some points on some flow streamlines. In this figure the orientation distribution is directly depicted on the unit surface, and the colors indicate the intensity of the orientation in each direction.

2.4 Conclusions

In this chapter we revisited different physical descriptions at different scales involved in material science. At the lowest scale, quantum mechanics involves high-dimensional models whose solution, when fermions are considered, must be anti-symmetric. We have pointed out that in fact, more than the multidimensional character of quantum mechanics models, the real challenge is the antisymmetry constraint.

Coarser modeling involves molecular dynamics or Brownian dynamics but despite its conceptual simplicity its use is not exempted from computational and conceptual difficulties. The next level is the one that corresponds to statistical mechanics descriptions leading to continuous models described by non-linear and coupled multidimensional partial differential equations. Some of these models have been successfully solved in some of our former works by using the adaptive separated representation within the PGD framework. This technique seems a promising alternative for addressing highly-multidimensional models, but as argued in the previous section, its use in quantum mechanics systems composed of fermions is limited, at present, by the antisymmetry restriction.

References

1. E. Cancès, M. Defranceschi, W. Kutzelnigg, C. Le Bris, Y. Maday, in *Computational Quantum Chemistry: a Primer*. Handbook of Numerical Analysis, vol 10 (Elsevier, Amsterdam, 2003), p. 3–270
2. C. Le Bris (ed.), *Handbook of Numerical Analysis*. Computational Chemistry, Vol. 10 (Elsevier, Amsterdam, 2003)
3. V. Gavini, K. Bhattacharya, M. Ortiz, Quasi-continuum orbital-free density-functional theory: a route to multi-million atom non-periodic DFT calculation. J. Mech. Phys. Solids **55**(4), 697–718 (2007)
4. C. Le Bris, P.L. Lions, From atoms to crystals: a mathematical journey. Bull. Am. Math. Soc. **42**(3), 291–363 (2005)
5. C. Le Bris, Computational chemistry from the perspective of numerical analysis. Acta Numerica **14**, 363–444 (2005)
6. C. Le Bris, in *Mathematical and Numerical Analysis for Molecular Simulation: Accomplishments and Challenges*. Proceedings of the International Congress of Mathematicians, (Madrid, Spain, 2003), pp. 1507–1522
7. V.B. Shenoy, R. Millera, E.B. Tadmor, D. Rodney, R. Phillips, M. Ortiz, An adaptive fnite element approach to atomic-scale mechanics—the quasicontinuum method. J. Mech. Phys. Solids **36**, 500–531 (1999)
8. G.J. Wagner, W.K. Liu, Coupling of atomistic and continuum simulations using a bridging scale decomposition. J. Comput. Phys. **190**, 249–274 (2003)
9. H. Ben Dhia, Multiscale mechanical problems: the Arlequin method. C. R. Acad. Sci., Paris 326, Ser-II b. **326**, 899–904 (1998)
10. S.P. Xiao, T. Belytschko, A bridging domain method for coupling continua with molecular dynamics. Comput. Methods Appl. Mech. Eng. **193**, 1645–1669 (2004)
11. H.C. Öttinger, M. Laso, in *Smart Polymers in Finite Element Calculation*, ed. by P. Moldanaers, R. Keunings. Theoretical and Applied Rheology. Proceedings of the XIth International Congress on Rheology, vol 1 (Elsevier, 1992), pp. 286–288

12. M.A. Martinez, E. Cueto, I. Alfaro, M. Doblare, F. Chinesta, Updated Lagrangian free surface flow simulations with the natural neighbour Galerkin methods. Int. J. Numer. Meth. Eng. **60**(13), 2105–2129 (2004)
13. D. Gonzalez, E. Cueto, F. Chinesta, M. Doblare, An updated Lagrangian strategy for free-surface fluid dynamics. J. Comput. Phys. **223**, 127–150 (2007)
14. M.A. Martinez, E. Cueto, M. Doblare, F. Chinesta, Natural element meshless simulation of injection processes involving short fiber suspensions. J. Nonnewton. Fluid Mech. **115**, 51–78 (2003)
15. A. Ammar, B. Mokdad, F. Chinesta, R. Keunings, A new family of solvers for some classes of multidimensional partial differential equations encountered in kinetic theory modeling of complex fluids. J. Nonnewton. Fluid Mech. **139**, 153–176 (2006)
16. F. Chinesta, A. Ammar, A. Falco, M. Laso, On the reduction of stochastic kinetic theory models of complex fluids. Model. Simul. Mater. Sci. Eng. **15**, 639–652 (2007)
17. B.B. Bird, C.F. Curtiss, R.C. Armstrong, O. Hassager, in *Dynamics of Polymeric Liquids*. Kinetic Theory, vol 2 (John Wiley and Sons, Chichester, 1987)
18. R.G. Larson, *The Structure and Rheology of Complex Fluids* (Oxford University Press, New York, 1999)
19. H.J. Bungartz, M. Griebel, Sparse grids. Acta Numerica **13**, 1–123 (2004)
20. Y. Achdou, O. Pironneau, *Computational Methods for Option Pricing, SIAM Frontiers in Applied Mathematics* (SIAM Publishers, Philadelphia, 2005)
21. A. Ammar, F. Chinesta, in *Circumventing Curse of Dimensionality in the Solution of Highly Multidimensional Models Encountered in Quantum Mechanics Using Meshfree Separated Representations*. Lecture Notes on Computational Science and Engineering, (Springer, 2008), pp. 1–17

Chapter 3
PGD-Based Computational Homogenization

Calculemus! (Let us calculate!)

—Gottfried Wilhelm Leibniz

Homogenization approaches are now intensively used in numerous engineering applications. These approaches belong to the more general category of multiscale methods, in which there are usually distinguished two main families of approaches, namely hierarchical and concurrent, see e.g. [1–3].

In hierarchical approaches, the behavior at the higher—macro—scale is obtained from the solution of a boundary value problem solved at the lower—micro—scale and posed on a representative volume element (RVE) of the material. This behavior can be determined under the form of macroscopic constitutive relations whose effective properties are identified from the solution of the microscopic problem. Micromechanical approaches fall in this category [4, 5]. The microscopic solution can also be incorporated in the macroscopic problem through a numerical scheme. It can be condensed at the macroscale as in the case of the variational multiscale method [6], or within a decomposition domain framework for structural mechanics applications [32]. The microscopic solution may also enrich the macroscale solution using strategies such as X-FEM and G-FEM, see e.g. [7] and [8] respectively. One of the major features of hierarchical approaches is that the macroscopic and microscopic problems are not coupled. This means that the microscopic problems can be solved once and for all, and consequently that only a finite number of microscopic solutions have to be known.

By contrast, the microscopic and macroscopic scales are solved simultaneously in concurrent approaches. This is required when the macroscopic behavior can not be obtained explicitly, which occurs as soon as microscopic nonlinear behaviors are involved. The area where the microscopic description is needed can be isolated but in this paper we will focus on problems where the whole domain is characterized by an underlying microstructure exhibiting a nonlinear behavior. These methods

F. Chinesta and E. Cueto, *PGD-Based Modeling of Materials, Structures and Processes*,
ESAFORM Bookseries on Material Forming, DOI: 10.1007/978-3-319-06182-5_3,

were reviewed in [33] in the continuum mechanics framework. The most popular approach is the FE^2 method initiated by Feyel [9], which has also become to be known as computational homogenization, see e.g. [10]. This method assumes that the microstructure is very small compared to the macroscopic domain. So at the macroscopic scale, it plays the role of a material point—an integration point if the finite element is used—at which an RVE is attached.

As noticed in [3], both the hierarchical and concurrent simulation approaches offer cost savings over large scale direct numerical simulation of the microstructure. However efforts are directed towards improving their numerical efficiency.

In hierarchical multiscale methods, one current challenge is the analysis of a representative volume element with a highly complex geometry. Powerful techniques such as computer tomography actually enables us to obtain high resolution 3D images of material microstructures. Thus an image with $1024 \times 1024 \times 1024$ voxels is now common, but its incorporation in a numerical model requires the use of an appropriate solution method.

In concurrent multiscale methods, such as the FE^2 approach, one main limitation is its computational cost, since several microscopic problems have to be solved at each integration point of the macroscopic domain. Therefore attempts have been made to build hierarchical approaches for nonlinear materials. One can thus define approximate nonlinear constitutive relations, see e.g. [11]. Another possibility is to build a discrete material map (see Ref. [12–14] for the case of nonlinear elasticity). This approach has been used in these references in nonlinear elasticity. It consists in discretizing the macroscopic strain space and determinating the microscopic solution for each node of this space. Even if the number of microscopic problems to be solved is large, this method is much more efficient than the FE^2 approach.

Recently, the authors have proposed a solution method based on the use of Proper Generalized Decomposition. In the field of homogenization, two simpler scenarios can be considered: (i) To define a generic microstructure from a voxel description and assume the conductivity of each voxel as an extra-coordinate; and (ii) To address nonlinear homogenization by solving the model at the microscopic level for any macroscopic state or history. This chapter has as its main objective opening some routes for addressing both scenarios thanks to the use of Proper Generalized Decompositions. For the sake of simplicity, and because this work is a first attempt in this ambitious objective, we will focus on the homogenization of linear and nonlinear (time-independent and time-dependent) thermal models. We are aware that this contribution is opening many routes without closing any of them, but as just mentioned this is not more than a preliminary analysis of the potential application of PGD for some multidimensional descriptions related to computational homogenization. The application of these ideas to more realistic scenarios as well as to thermomechanical models is a work in progress.

3.1 Homogenization of Linear Heterogenous Models

3.1.1 Formulation of the Homogenization Problem

Due to the microscopic heterogeneity, the macroscopic thermal modeling needs an homogenized thermal conductivity which depends on the microscopic details.

To compute this homogenized thermal conductivity in the linear case, one could isolate a representative volume element at position $\mathbf{X}$, $\Omega_{rve}(\mathbf{X})$ (this representative volume element is supposed to be defined, using for example methods which are proposed in [15–17]) and assume that the microstructure is perfectly defined at such scale. Thus, the microscopic conductivity $\mathbf{k}(\mathbf{x})$ is known at each point in the microscopic domain $\mathbf{x} \in \Omega_{rve}(\mathbf{X})$.

We can define the macroscopic temperature gradient at position $\mathbf{X}$, $\mathbf{G}(\mathbf{X})$, from:

$$\mathbf{G}(\mathbf{X}) = \langle \mathbf{g} \rangle = \frac{1}{|\Omega_{rve}(\mathbf{X})|} \int_{\Omega_{rve}(\mathbf{X})} \mathbf{g}(\mathbf{x})\, d\Omega, \tag{3.1}$$

where $\mathbf{g}(\mathbf{x}) = \nabla T(\mathbf{x})$.

We also assume the existence of a localization tensor $\mathbf{L}(\mathbf{x}, \mathbf{X})$ such that

$$\mathbf{g}(\mathbf{x}) = \mathbf{L}(\mathbf{x}, \mathbf{X}) \cdot \mathbf{G}(\mathbf{X}). \tag{3.2}$$

Now, we consider the microscopic heat flux $\mathbf{q}$ according to Fourier law

$$\mathbf{q}(\mathbf{x}) = -\mathbf{k}(\mathbf{x}) \cdot \mathbf{g}(\mathbf{x}), \tag{3.3}$$

and the macroscopic counterpart $\mathbf{Q}(\mathbf{X})$ that is written:

$$\mathbf{Q}(\mathbf{X}) = \langle \mathbf{q}(\mathbf{x}) \rangle = \langle -\mathbf{k}(\mathbf{x}) \cdot \mathbf{g}(\mathbf{x}) \rangle = \langle -\mathbf{k}(\mathbf{x}) \cdot \mathbf{L}(\mathbf{x}, \mathbf{X}) \rangle \cdot \mathbf{G}(\mathbf{X}), \tag{3.4}$$

from which the homogenized thermal conductivity can be defined from

$$\mathbf{K}(\mathbf{X}) = \langle -\mathbf{k}(\mathbf{x}) \cdot \mathbf{L}(\mathbf{x}, \mathbf{X}) \rangle. \tag{3.5}$$

Since $\mathbf{k}(\mathbf{x})$ is perfectly known everywhere in the representative volume element, the definition of the homogenized thermal conductivity tensor only requires the computation of the localization tensor $\mathbf{L}(\mathbf{x}, \mathbf{X})$. Several approaches are proposed in the literature to define this tensor, according to the choice of the boundary conditions. Our objective here is not to discuss this choice. The interested reader can find some details in [15, 18, 19], or [20]. For the sake of simplicity, we use essential boundary conditions corresponding to the assumption of uniform temperature gradient on the RVE boundary. Hence we are led to consider in the 3D case the solution of the three boundary value problems related to the steady state heat transfer model in the microscopic domain, defined by:

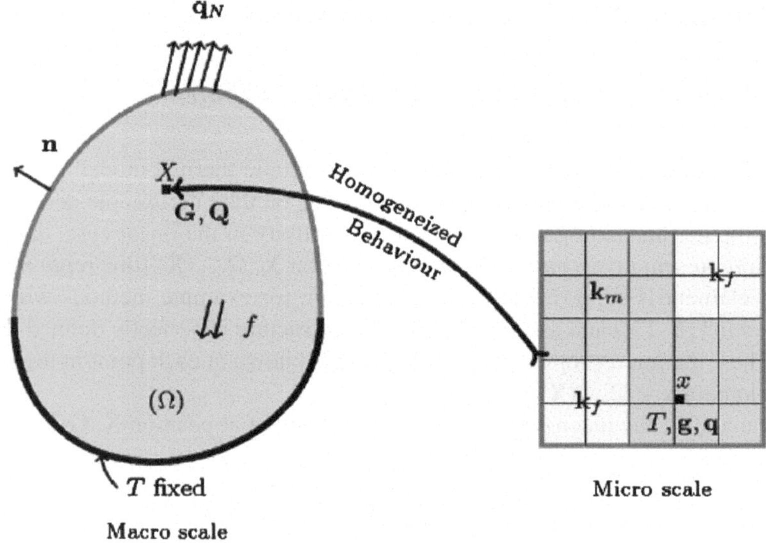

Fig. 3.1 Homogenization procedure of linear heterogenous models

$$\begin{cases} \nabla \cdot \big(\mathbf{k}(\mathbf{x}) \cdot \nabla T^1(\mathbf{x})\big) = 0 \\ T^1(\mathbf{x} \in \partial\Omega_{rve}) = x \end{cases}, \tag{3.6}$$

$$\begin{cases} \nabla \cdot \big(\mathbf{k}(\mathbf{x}) \cdot \nabla T^2(\mathbf{x})\big) = 0 \\ T^2(\mathbf{x} \in \partial\Omega_{rve}) = y \end{cases} \tag{3.7}$$

and

$$\begin{cases} \nabla \cdot \big(\mathbf{k}(\mathbf{x}) \cdot \nabla T^3(\mathbf{x})\big) = 0 \\ T^3(\mathbf{x} \in \partial\Omega_{rve}) = z \end{cases}. \tag{3.8}$$

It is easy to prove that these three solutions verify

$$\begin{cases} \mathbf{G}^1 = \langle \nabla T^1(\mathbf{x})\rangle^T = (1, 0, 0) \\ \mathbf{G}^2 = \langle \nabla T^2(\mathbf{x})\rangle^T = (0, 1, 0) \\ \mathbf{G}^3 = \langle \nabla T^3(\mathbf{x})\rangle^T = (0, 0, 1) \end{cases} \tag{3.9}$$

where $(\cdot)^T$ denotes the transpose. Thus, the localization tensor results finally:

$$\mathbf{L}(\mathbf{x}, \mathbf{X}) = \Big(\nabla T^1(\mathbf{x}) \quad \nabla T^2(\mathbf{x}) \quad \nabla T^3(\mathbf{x})\Big). \tag{3.10}$$

The resulting non-concurrent homogenization procedure is illustrated in Fig. 3.1. As soon as tensor $\mathbf{L}(\mathbf{x}, \mathbf{X})$ is known at each position $\mathbf{X}$, the constitutive law relating the macroscopic temperature gradient and macroscopic heat flux becomes defined.

From the computational point of view the main difficulty concerns the solution of the three boundary value problems using very fine meshes required to represent accurately all the microscopic details as well as the application of this procedure at different places for accounting for the spatial evolution of the microstructure of its stochastic nature.

3.1.2 Separated Representation of the Physical Space

As mentioned in the introduction, recent advances in imaging techniques offer the opportunity to obtain high resolution images of material microstructures. This leads to very large finite element models which can reach one billion degrees of freedom, see e.g. [21]. Nowadays, the performance of computers fails to solve, in a reasonable time and with classical solvers, problems of this size. In order to circumvent this difficulty we propose to use a separated representation of the physical space.

Thus, temperature fields involved in the thermal models (3.6–3.8) are searched in a separated form. Therefore, at iteration n the solution is written

$$T^n(x, y, z) = \sum_{i=1}^{i=n} X_i(x) \cdot Y_i(y) \cdot Z_i(z), \tag{3.11}$$

and it is enriched until reaching convergence by computing the next term in the finite sums decomposition

$$T^{n+1}(x, y, z) = T^n(x, y, z) + R(x) \cdot S(y) \cdot W(z). \tag{3.12}$$

Applying the strategy described in the previous section we compute the function $R(x)$ assuming known $S(y)$ and $W(z)$, then $S(y)$ from the just computed $R(x)$ and $W(z)$ and finally $W(z)$ from $R(x)$ and $S(x)$. After reaching convergence we can assign $R(x) \rightarrow X_{n+1}(x)$, $S(y) \rightarrow Y_{n+1}(y)$ and $W(z) \rightarrow Z_{n+1}(z)$.

We assume that the microscopic conductivity can be written in a separated form, i.e. each component of the conductivity tensor $\mathbf{k}_{ij}(\mathbf{x})$ can be expressed from:

$$\mathbf{k}_{ij}(\mathbf{x}) \approx \sum_{p=1}^{p=M} (\mathbf{k}_{ij}^1)_p(x) \cdot (\mathbf{k}_{ij}^2)_p(y) \cdot (\mathbf{k}_{ij}^3)_p(z) \tag{3.13}$$

this representation can be performed by applying for example a multidimensional singular value decomposition (SVD) [22]. Following the procedure previously described, the computation of $R(x)$ implies the solution of the second order BVP

$$\alpha_x \frac{d^2R(x)}{dx^2} + \beta_x \frac{dR(x)}{dx} + \gamma_x R(x) = f_x(x), \tag{3.14}$$

where the coefficients depend on the integrals of functions depending on y and z in their respective domains. Analogously, the calculation of functions $S(y)$ and $W(z)$ implies the solution of similar equations but implying different coefficients and source terms:

$$\alpha_y \frac{d^2S(y)}{dy^2} + \beta_y \frac{dS(y)}{dy} + \gamma_y S(y) = f_y(y), \tag{3.15}$$

and

$$\alpha_z \frac{d^2W(z)}{dz^2} + \beta_z \frac{dW(z)}{dz} + \gamma_z W(z) = f_z(z). \tag{3.16}$$

Thus, as proved in [23], models involving millions of nodes along each direction can be solved because the procedure based on separated representations avoids the definition of a mesh. This enables degrees of resolution never until now reached, the main issue being the solution of the one dimensional boundary value problems. In general the discrete form of these problems leads to tridiagonal matrices that can be easily solved. In the next section we are describing a procedure, originally proposed in [24], specially adapted to such kind of solutions that avoids the solution of a large size linear system.

Remark 3.1.1 The main limitation in using this fully separated representation is precisely the necessity to build-up a separated approximation of the microscopic thermal conductivity (3.13) because complex microstructures imply thousands of terms in the sum (3.13) as well as the necessity of defining the conductivity in a grid of the representative volume element prior to applying a singular value decomposition -SVD-. The separated representation of the conductivity at the microscopic level is not mandatory, but without such a representation, integrals in the whole domain have to be computed, which can dramatically reduce the computational efficiency. In Sect. 3.3 we explore some possibilities for alleviating this drawback.

3.1.3 Efficient Solvers of One-Dimensional BVP

Consider the generic second order BVP implying the field $u(x)$, $x \in [0, L]$

$$a \frac{d^2u}{dx^2} + b \frac{du}{dx} + cu = f(x) \tag{3.17}$$

with $u(x = 0) = u_l$ and $u(x = L) = u_r$.

The solution of this equation can be written as the addition of the general solution of the homogeneous equation and a particular solution of the complete equation, that is denoted by $h^h(x)$ and $u^c(x)$ respectively.

The solution $u^c(x)$ can be computed by integrating equation

$$a\frac{d^2u^c}{dx^2} + b\frac{du^c}{dx} + cu^c = f(x) \tag{3.18}$$

from $u^c(x=0) = 0$ and $\left.\frac{du^c}{dx}\right|_{x=0} = 0$. The integration can be performed by using a standard backward finite difference scheme, avoiding the solution of any linear system.

Now the solution of the homogeneous equation can be expressed as $u^h(x) = \alpha u_1^h(x) + \beta u_2^h(x)$, where $u_1^h(x)$ and $u_2^h(x)$ results from the backward integration of

$$a\frac{d^2u_1^h}{dx^2} + b\frac{du_1^h}{dx} + cu_1^h = 0, \tag{3.19}$$

with $u_1^h(x=0) = 0$ and $\left.\frac{du_1^h}{dx}\right|_{x=0} = 1$; and

$$a\frac{d^2u_2^h}{dx^2} + b\frac{du_2^h}{dx} + cu_2^h = 0, \tag{3.20}$$

with $u_2^h(x=0) = 1$ and $\left.\frac{du_2^h}{dx}\right|_{x=0} = 0$.

Thus, finally, the general solution is written $u(x) = u^c(x) + \alpha u_1^h(x) + \beta u_2^h(x)$. The coefficients α and β can be easily computed by enforcing the model boundary conditions:

$$\beta = u_l, \tag{3.21}$$

because $u^c(x=0) = u_1^h(x=0) = 0$ and $u_2^h(x=0) = 1$, and

$$\alpha = \frac{u_r - u^c(x=L) - u_l \cdot u_2^h(x=L)}{u_1^h(x=L)}. \tag{3.22}$$

The advantage of such procedure lies in its simplicity to be implemented in any computing platform even when considering thousands of millions of discrete values [24–26].

3.1.4 Enriched Parametric Models

Many materials exhibit a random microstructure, needing the solutions of different realizations of the microstructure. Such situation is also encountered for a numerical determination of the RVE size, see e.g. [13, 15, 16]. Other times the microstructure is known everywhere, but as it could evolve in the macroscopic scale, one should solve a homogenization problem at each integration point (or nodal position) at the macroscopic scale. In this section we are describing an efficient way of circumventing this computational complexity.

For the sake of simplicity, a two-dimensional thermal model defined in $\Omega \subset \Re^2$ is considered. It is characterized by a microstructure that evolves in the macroscopic scale ($\mathbf{X}$). Thus, at each position $\mathbf{X}$, a representative volume element $\Omega_{rve}(\mathbf{X}) = [0, L]^2$ can be extracted.

In the following, points in $\Omega_{rve}(\mathbf{X})$ are denoted by $\mathbf{x}$. We assume that the microstructure in each $\Omega_{rve}(\mathbf{X})$ consists of $M \times M$ cells, each one characterized by a value of the thermal conductivity.

If $h = \frac{L}{M}$, the cells boundaries are defined by the segments $(x_i = (i-1) \times h, y)$, $i = 1, \ldots, M+1$ and $y \in [0, L]$; and $(x, y_i = (i-1) \times h)$, $i = 1, \ldots, M+1$ and $x \in [0, L]$.

If we define the characteristic function $\chi_k^x(x)$, $k = 1, \ldots, M$ by

$$\chi_k^x(x) = \begin{cases} 1 & \text{if } x_k < x < x_{k+1} \\ 0 & \text{otherwise} \end{cases}, \tag{3.23}$$

and analogously $\chi_k^y(y)$, $k = 1, \ldots, M$ by

$$\chi_k^y(y) = \begin{cases} 1 & \text{if } y_k < y < y_{k+1} \\ 0 & \text{otherwise} \end{cases}, \tag{3.24}$$

any cell C_{ij} with $i, j = 1, \ldots M$, can be defined by $C_{ij} = \chi_i^x(x) \times \chi_j^y(y)$.

Using this notation the thermal conductivity at the microscopic level can be expressed in the separated form:

$$\mathbf{k}(\mathbf{x}) = \sum_{i=1}^{i=M} \sum_{j=1}^{j=M} \mathbf{k}_{ij} \, \chi_i^x(x) \times \chi_j^y(y). \tag{3.25}$$

In what follows for the sake of simplicity we are assuming that the thermal conductivity is isotropic in each cell, assumption that allows writing

$$k(\mathbf{x}) = \sum_{i=1}^{i=M} \sum_{j=1}^{j=M} k_{ij} \, \chi_i^x(x) \times \chi_j^y(y). \tag{3.26}$$

Now, two possibilities can be considered to perform homogenization:

1. The simplest one consists in applying the procedure described in Sect. 3.1. Thus, three steady state thermal models should be solved at each point $\mathbf{X}$ on the corresponding microscopic domain $\Omega_{rve}(\mathbf{X})$ where the homogenized thermal properties are searched.

2. The second possibility consists in introducing the cell thermal conductivities as extra-coordinates of the thermal model, which implies the following separated form of the temperature field:

$$T(x, y, k_{1,1}, k_{1,2}, \ldots, k_{M,M}) \approx \sum_{l=1}^{l=N} X_l(x) \cdot Y_l(y) \cdot \prod_{p=1}^{p=M} \prod_{q=1}^{q=M} \Xi^l_{pq}(k_{p,q}). \tag{3.27}$$

By solving the resulting thermal model only once, but defined in a space involving $2 + M \times M$ dimensions, one could have direct access to the temperature at each point for any value of the thermal conductivities in each cell composing the microscopic representative element volume. Consequently, when we move from one position $\mathbf{X}$ to another one $\mathbf{X}'$, as soon as the different cell thermal conductivities are known at this new position $\mathbf{X}'$, the temperature field is determined from Eq. (3.27). This allows the computation of the homogenized thermal conductivity without the necessity of solving a thermal model at each location in the macroscopic domain. The price to be paid is the solution of a thermal model in a multidimensional space, but this solution, thanks to the proper generalized decomposition can be performed, only once and off-line.

Remark 3.1.2 As just argued in Sect. 3.2, the fact that the thermal model does not contain derivatives with respect to the thermal conductivity, implies that the introduction of the different cell conductivities as extra-coordinates has not a significant impact in the computational efforts needed to construct the separated representation. However, the number of terms involved in the separated representation, N, can increase significantly with the number of extra-coordinates introduced in the model. In [27] the solution of models containing hundreds of dimensions was performed in some minutes using a standard laptop, however, the solution of the models just described needed hundreds of terms and computations ranging from some hours to some days. In any case, as indicated previously, this multidimensional problem should be solved only once and off-line.

3.2 Advanced Non-concurrent Non-linear Homogenization

Until now only linear behaviors were addressed, however in practice the most serious difficulties appear as soon as nonlinear behaviors are considered. In this section, we focus on this issue, and more precisely in two scenarios: The first one concerns nonlinear behaviors independent on the thermal history, and the second one consider the more general case in which the thermal properties at a certain point and time depends not only on the present temperature, but also on its entire thermal history.

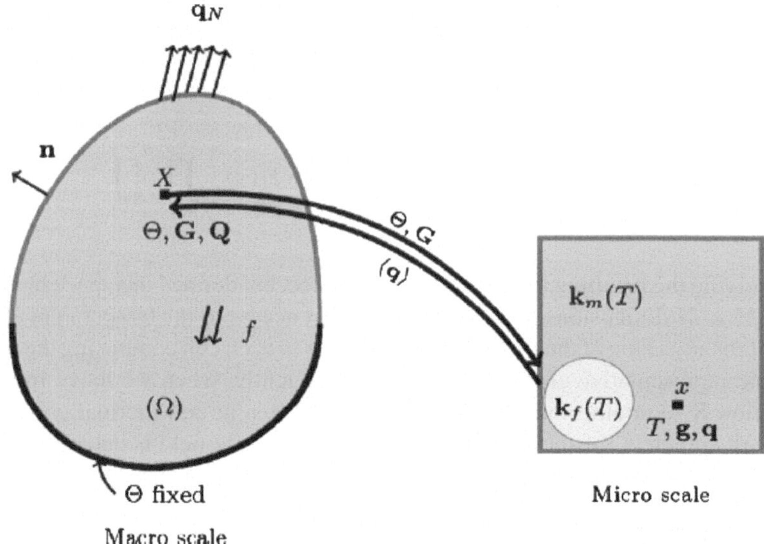

Fig. 3.2 Concurrent homogenization of non-linear history independent thermal models

3.2.1 History Independent Non-linear Behaviors

Using the notation introduced previously we assume that in a certain representative volume element $\Omega_{rve}(\mathbf{X})$ located at position $\mathbf{X}$, the temperature gradient $\mathbf{G}(\mathbf{X})$ is known (from the previous homogenized macroscopic thermal model solution). We are interested in knowing the associated heat flux $\mathbf{Q}(\mathbf{X})$ at that position. In the present case, the nonlinearity is coming from the fact that the thermal conductivity depends on the temperature itself, i.e. $\mathbf{k}(T)$.

Usual concurrent homogenization proceeds as illustrated in Fig. 3.2. The macroscopic temperature and the macroscopic temperature gradient are used for computing at the microscopic level the averaged heat flux that corresponds to the macroscopic heat flux.

If one solves the thermal model in $\Omega_{rve}(\mathbf{X})$

$$\begin{cases} \nabla \cdot (\mathbf{k}(T) \cdot \nabla T(\mathbf{x})) \ = \ 0 \\ T(\mathbf{x} \in \partial\Omega_{rve}) = \mathbf{G} \cdot \mathbf{x} \end{cases} \tag{3.28}$$

it is easy to prove that $\langle \nabla T(\mathbf{x}) \rangle = \mathbf{G}$.

The main difficulty lies in the fact that as soon as the gradient of temperature changes at position $\mathbf{X}$ a new thermal model must be solved in the associated representative volume element $\Omega_{rve}(\mathbf{X})$. In the linear case, only a 3D thermal model for each space coordinate must be solved in each representative volume element

$\Omega_{rve}(\mathbf{X})$. But now, as just argued, the complexity increases because of the dependence of the homogenized thermal properties on the thermal state and not only on the microstructure.

To simplify the solution process, the thermal model in $\Omega_{rve}(\mathbf{X})$ could be solved for any macroscopic gradient $\mathbf{G}$. Thus, as soon as the macroscopic gradient is known, the macroscopic heat flux $\mathbf{Q}$ could be obtained without the necessity of solving a new thermal model. The main idea lies in introducing the components of $\mathbf{G}$ (denoted by $\mathbf{G}_1, \mathbf{G}_2, \mathbf{G}_3$) as extra-coordinates in the thermal model.

The main issue is how to transfer $\mathbf{G}$ from the boundary conditions to the partial differential equation itself. For this purpose, a new temperature field $\theta(\mathbf{x})$ is defined as:

$$\theta(\mathbf{x}) = T(\mathbf{x}) - \mathbf{G} \cdot \mathbf{x}. \tag{3.29}$$

The thermal model (3.28) becomes:

$$\begin{cases} \nabla \cdot (\mathbf{k}(\theta, \mathbf{G}) \cdot \nabla(\theta(\mathbf{x}) + \mathbf{G} \cdot \mathbf{x})) = 0 \\ \theta(\mathbf{x} \in \partial\Omega_{rve}) = 0 \end{cases} \tag{3.30}$$

whose solution is sought in a separated form

$$\theta(x, y, z, \mathbf{G}_1, \mathbf{G}_2, \mathbf{G}_3) \approx \sum_{i=1}^{i=N} X_i(x) \cdot Y_i(y) \cdot Z_i(z) \cdot \Gamma_i^1(\mathbf{G}_1) \cdot \Gamma_i^2(\mathbf{G}_2) \cdot \Gamma_i^3(\mathbf{G}_3), \tag{3.31}$$

by applying the proper generalized decomposition previously described.

In mechanics, (for example nonlinear elasticity), this procedure suffices for determining the macroscopic stress ■ from the macroscopic deformation tensor $\mathbf{E}$. This approach is sufficient because prescribing a displacement field $\mathbf{u} = \mathbf{E} \cdot \mathbf{x}$ on the boundary of Ω_{rve} ensures that the resulting microscopic displacement field verifies $\langle \varepsilon \rangle = \mathbf{E}$. Obviously, a displacement given by $\mathbf{E} \cdot \mathbf{x} + \mathbf{U}_0$ (with $\mathbf{U}_0$ a uniform translation) could be prescribed on the boundary without modifying the solution because $\langle \varepsilon \rangle = \mathbf{E}$ is still satisfied as the elastic constants depend on the strain but not on the displacement. In mechanical models only the gradient of the displacement plays a role. Thus, if a new displacement field is defined as:

$$\widetilde{\mathbf{u}}(\mathbf{x}) = \mathbf{u}(\mathbf{x}) - \mathbf{E} \cdot \mathbf{x} - \mathbf{U}_0. \tag{3.32}$$

The introduction of the resulting expression of $\mathbf{u}(\mathbf{x})$ into the nonlinear elastic problem, would make the translation field $\mathbf{U}_0$ disappears because (i) it is constant and then it vanishes when derivatives apply, and (ii) the elastic coefficients only depends on the gradient of the displacement (strain) and not on the displacement itself.

Other models however exhibit such a dependence on the field itself. For example in thermal models, Fourier's law relates the temperature gradient with the heat flux through the material conductivity that usually depends on the temperature itself.

The same situation can be found in the Fick's diffusion law relating the concentration gradient and the chemical species fluxes through the so-called diffusion coefficient that is usually assumed depending on the concentration itself. In all these cases the procedure just described for computing the macroscopic flux from the macroscopic gradient (in our case $\mathbf{Q}$ from $\mathbf{G}$) fails: Prescribing the temperature $\mathbf{G} \cdot \mathbf{x} + \mathbf{T}_0$ on the boundary of Ω_{rve} ensures condition $\langle \nabla T \rangle = \mathbf{G}$. However, the validity of the the computed $\mathbf{Q}$ cannot be guaranteed because the computed temperature field depends on T_0, due to the dependence of the thermal conductivity on the temperature itself.

This problem of transferring the macroscopic temperature to the microscopic problem is addressed in [20]. In this reference, the microscopic temperature over the RVE is linked to the macroscopic temperature through an equation which expresses the consistency of the stored heat at the macro and the micro level.

So we can see that as in [20], the data of the macroscopic problem are composed with the macroscopic temperature and the gradient of temperature. Therefore, the temperature field is searched under the following form, after applying the change of variable for transferring the boundary condition to the equation itself:

$$\theta(x, y, z, \mathbf{G_1}, \mathbf{G_2}, \mathbf{G_3}, \mathbf{T_0}) \approx \sum_{i=1}^{i=N} X_i(x) \cdot Y_i(y) \cdot Z_i(z) \cdot \Gamma_i^1(\mathbf{G}_1) \cdot \Gamma_i^2(\mathbf{G}_2) \cdot \Gamma_i^3(\mathbf{G}_3) \cdot \Upsilon_i(T_0), \tag{3.33}$$

3.2.2 History Independent Non-linear Behaviors: A Simplified Procedure

We have just proven that the accurate homogenization of nonlinear behaviors, in absence of internal variables (and then in absence of history dependencies), needs to transfer two key pieces of information from the macroscopic scale to the microscopic one: (i) The gradient of the macroscopic temperature and (ii) The macroscopic temperature itself. The resulting microscopic problem is nonlinear (the thermal properties depend on the microscopic temperature) and as just described, one equation links the microscopic temperature to the macroscopic one.

Even if the procedure is conceptually simple, its numerical implementation deserves certain difficulties since the solution is nonlinear with respect to the gradient of the macroscopic temperature. Thus, many authors proposed a simplified modeling in which the thermal properties at the microscopic level (that depend on the existing local temperature) are frozen to values corresponding to the macroscopic temperature. That is, the thermal conductivity at any point in the microscopic RVE is fixed according to the macroscopic temperature, and then it does not evolve with the local microscopic temperature. This approach has been rigorously justified from the asymptotic expansion method for materials with periodic microstructure in [28]. In a few words, in this paper, the authors have shown that for a given

0-th order temperature—i.e. macroscopic temperature—the effective conductivity of the non-linear temperature dependent problem is equal to that of the linear problem with thermal properties fixed at the macroscopic temperature. Therefore this paper provides a justification of an approach which was widely applied in the context of computational homogenization of non-linear thermal models, see e.g. [29–31].

Now, with the thermal properties frozen at the microscopic scale, the non-linearity of the microscopic model disappears. Thus, the microscopic solution only needs the prescription of the macroscopic temperature gradient $\mathbf{G}(\mathbf{X})$ as soon as the thermal conductivity is assumed known everywhere in the microscopic representative volume.

The simplest possibility for computing a general solution is solving the microscopic linear models defined in Sect. 3.1.1. (two in the 2D case and 3 in the 3D case) for the material thermal conductivity related to any macroscopic temperature $\Theta(\mathbf{X})$. The solution of these microscopic problems ($T^1(\mathbf{x})$, $T^2(\mathbf{x})$ and $T^3(\mathbf{x})$ in Sect. 3.1.1) for any macroscopic temperature Θ suggests the separated representation:

$$T^j(x, y, z, \Theta) \approx \sum_{i=1}^{i=N} X_i(x) \cdot Y_i(y) \cdot Z_i(z) \cdot \theta_i(\Theta); \quad j = 1, 2, 3, \tag{3.34}$$

These solutions allow computing the homogenized conductivity tensor for any value of the macroscopic temperature Θ. Through the dependence of the conductivity tensor to the macroscopic temperature, this homogenization problem displays similarities with homogenization in nonlinear elasticity, for which non-concurrent multiscale approaches have been recently proposed [12–14]. In these references, the macroscopic nonlinear behavior is computed after the discretization of the macroscopic strain space. Then, in a second step, a continuous representation is determined by an interpolation technique. It can be seen that our approach provides automatically this interpolation and therefore enables to compute in an elegant and efficient way the homogenized behavior of nonlinear models.

Remark 3.2.1 The gradient of the temperature fields given by Eq. (3.34) allows computing the homogenized macroscopic conductivity tensor within a non-concurrent framework. However, because of the nonlinearity of the macroscopic homogenized thermal model (i.e., the macroscopic homogenized thermal conductivity depends on the macroscopic temperature) we should determine an expression of the macroscopic tangent conductivity. This can be easily obtained from the derivatives of the different components of the homogenized conductivity tensor with respect to the macroscopic temperature. From Eq. (3.34) we can actually notice that for this purpose it suffices to take the derivative of the spatial gradient of the microscopic temperatures ∇T^1, ∇T^2 and ∇T^3 involved in the definition of K(X) with respect to the macroscopic temperature Θ. Obviously, this procedure is very easy to use, whereas in [12, 14] the tangent operators are computed once the interpolation of the solutions obtained in the discretized macroscopic space is built.

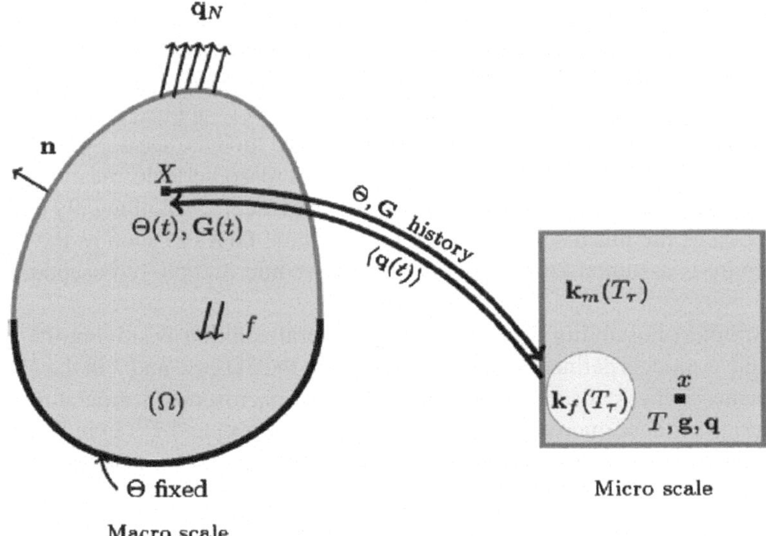

Fig. 3.3 Concurrent homogenization of nonlinear history dependent thermal models

3.2.3 History Dependent Non-linear Behaviors

In the previous section we assumed that the thermal model coefficient depends on the instantaneous temperature, but for the sake of generality we extend the previous procedure (which consists in introducing all the thermal loads as extra-coordinates) to models in which the thermal parameters also depend on the thermal history. This scenario is the equivalent of mechanical models involving internal variables, as elastoplasticity for example.

Usual concurrent homogenization proceeds as illustrated in Fig. 3.3. The macroscopic temperature and the macroscopic temperature gradient histories are used for computing at the microscopic level the averaged heat flux at present time that corresponds to the macroscopic heat flux.

In what follows, we assume, without looking for any physical interpretation, that the thermal conductivity at a certain location $\mathbf{x} \in \Omega_{rve}(\mathbf{X})$ depends on the thermal history at that position, i.e. $\mathbf{k} = \mathbf{k}(T(\tau)), 0 \leq \tau \leq t$.

Now, as in the previous case, we should solve the thermal model defined in $\Omega_{rve}(\mathbf{X})$, but in this case the thermal model must be integrated in the whole time interval $0 \leq t \leq t_{max}$:

$$\frac{\partial T}{\partial t} - \nabla \cdot (\mathbf{k}(T(t)) \cdot \nabla T) = 0, \tag{3.35}$$

with $T(\mathbf{x} \in \partial\Omega_{rve}, t) = \mathbf{G}(t) \cdot \mathbf{x} + T_0(t)$.

We would like to introduce the components of $\mathbf{G}(t)$ and $T_0(t)$ as extra-coordinates in the partial differential equation describing the thermal model. However, as these potential candidates become coordinates, which still depend on the time, they cannot be at present assumed as coordinates in the model (the model coordinates must be independent).

Let $\mathbf{G}^i(t)$, $i = 1, 2, 3$ be the components of vector $\mathbf{G}(t)$. We could approximate the evolution of each one of these components using a set of discrete values. The simplest choice consists of using a simple linear finite element interpolation:

$$\mathbf{G}^i(t) = \sum_{j=1}^{j=M} N_j(t) \cdot \mathbf{G}^i_j, \tag{3.36}$$

where $\mathbf{G}^i_j = \mathbf{G}^i(t = t_j)$, $t_j = (j-1) \times \Delta t$, and $N_j(t)$ the standard linear finite element shape functions.

We proceed in a similar way for approximating $T_0(t)$,

$$T_0(t) = \sum_{j=1}^{j=M} N_j(t) \cdot (T_0)_j. \tag{3.37}$$

Now, the discrete values $\mathbf{G}^i_j$, $i = 1, 2, 3$ and $j = 1, \ldots, M$; and $(T_0)_j$, $j = 1, \ldots, M$, all them being independent, can be considered as extra-coordinates in the transient thermal model. By solving it only once, and possibly off-line, we have direct access to the solution of the thermal model for any prescribed macroscopic history of the thermal load $\mathbf{G}(t)$ and $T_0(t)$. The price to be paid is the solution of a transient model in a space of dimension $3 + 1 + 4 \times M$.

As proposed earlier, the solution of the resulting multidimensional model is sought in a separated form (to allow a linearly scaling complexity with respect to dimensionality). Thus, the solution results in:

$$T(x, y, z, t, G^1_1, \ldots, G^1_M, \ldots, G^3_1, \ldots, G^3_M, (T_0)_1, \ldots, (T_0)_M) \approx$$

$$\sum_{k=1}^{k=N} X_i(x) \cdot Y_i(y) \cdot Z_i(z) \cdot \Psi_i(t) \cdot \left(\prod_{j=1}^{j=3} \prod_{k=1}^{k=M} \Gamma_i^{j,k}\left(\mathbf{G}^j_k\right) \right) \cdot \left(\prod_{k=1}^{k=M} \Upsilon_i^k\left((T_0)_k\right) \right). \tag{3.38}$$

Remark 3.2.2 As in the previous case, prior to introduce these extra-coordinates in the transient thermal model, they should be transferred from the boundary condition into the PDE itself. This transfer is performed as in the former case by applying the change of variable $\theta(\mathbf{x}, t) = T(\mathbf{x}, t) - \mathbf{G}(t) \cdot \mathbf{x} - T_0(t)$.

3.2.4 *History Dependent Non-linear Behaviors: A Simplified Procedure*

Even in the case of history dependent models, the complexity of the computation can be drastically reduced when considering the simplifying hypothesis proposed in Sect. 3.2.2 (thermal properties frozen to the values dictated by the macroscopic scale).

As just described, the proper solution of this kind of models needs the introduction of many time evolutions: Those related to each component of the macroscopic temperature gradient and the one related to the time evolution of T_0 that is prescribed on the domain boundary to enforce that the microscopic averaged temperature equals the macroscopic one at any time. Thus, the use of M values for describing the time evolution of these fields, implies the introduction, in the general 3D case, of $3 \times M + M$ extra-coordinates as just described in the previous section. However, if the thermal properties are assumed evolving as dictated by the macroscopic temperature evolution, the nonlinearity at the microscopic level disappears, and it suffices to compute the homogenized conductivity tensor as described in Sect. 4. For any evolution of the macroscopic temperature, i.e.:

$$T^j(x, y, z, t, \Theta_1, \ldots, \Theta_M) \approx \sum_{i=1}^{i=N} X_i(x) \cdot Y_i(y) \cdot Z_i(z) \cdot \Psi_i(t) \cdot \left(\prod_{k=1}^{k=M} \theta_i^k(\Theta_k) \right), \tag{3.39}$$

that reduces the dimensionality of the problem with respect to problem (3.38) by removing $3 \times M$ coordinates.

3.3 Numerical Examples

3.3.1 *Homogenization of Linear Models*

In Ammar et al. [23], the use of separated representations for computing the homogenized behavior of linear thermal models was already addressed. In this paper we retained the use of such technique (see Sect. 3.1.2), because it is used in Sect. 3.2 and in the numerical examples related to the procedures introduced in this section.

In what follows, we focus on the homogenization of materials whose microstructure can be accurately described by a microscopic representative volume element consisting of a certain number of cells, according to the description addressed in Sect. 3.1.4. For the sake of simplicity in what follows, and without loss of generality, only the 2D case is considered.

More precisely, we assume a 2D microscopic RVE $\Omega_{vre} = (0, 1) \times (0, 1)$ composed of 5×5 cells whose conductivity can take any value in the interval $[k_{\min}, k_{\max}]$, with $k_{\min} = 1$ and $k_{\max} = 2$ in the following.

According to the procedure described in Sect. 3.1.4, the linear thermal model is solved in Ω_{rve} for two different boundary conditions: $T(\mathbf{x} \in \partial\Omega_{rve}) = x$ in the first case and $T(\mathbf{x} \in \partial\Omega_{rve}) = y$ in the second case. The resulting solutions are denoted $T^1(\mathbf{x})$ and $T^2(\mathbf{x})$ respectively. As described in Sect. 3.1.2 these solutions suffice for defining the homogenized conductivity tensor. On the other hand, in order to compute solutions whose validity does not depend on the particular value of the conductivity of each cell, both solutions are computed by considering the thermal conductivities of all the cells as extra-coordinates, i.e. both solutions are sought under the form:

$$T^j(x, y, k_{1,1}, k_{1,2}, \ldots, k_{5,5}) \approx \sum_{l=1}^{l=N} X_l(x) \cdot Y_l(y) \cdot \prod_{p=1}^{p=5} \prod_{q=1}^{q=5} \Xi^l_{pq}(k_{p,q}), \quad (3.40)$$

for $j = 1, 2$.

Thus, instead of solving two thermal models for each different microstructure, the solution of two multidimensional models (defined in a space of dimension $2 + 5 \times 5 = 27$) suffices to give access to the temperature field for any possible microstructure defined on this grid. Moreover, thanks to the PGD, the solution of these multidimensional models does not introduce major difficulties.

With these solutions computed, and as soon as the microstructure is given, (value of the conductivity in each cell), the homogenized conductivity tensor can be computed.

In order to validate the solutions, a particular microstructure is defined by taking random values of the components of the conductivity tensor k_{ij} that we denote by k^g_{ik} (see Fig. 3.4). Then, the solution is obtained by particularizing the general PGD solution

$$T^1(x, y) \approx \sum_{l=1}^{l=N} X_l(x) \cdot Y_l(y) \cdot \prod_{p=1}^{p=5} \prod_{q=1}^{q=5} \Xi^l_{pq}(k^g_{p,q}), \; T^1(\mathbf{x} \in \partial\Omega_{rve}) = x. \quad (3.41)$$

Temperatures $T^1(x, y)$ and $T^2(x, y)$ obtained using the PGD are compared with finite element solutions defined on the same microstructure. The PGD results were obtained by considering a 101 nodes F.E. discretization for functions of x and y and a three nodes discretization for the functions depending on the different conductivities. The classical F.E. solution was obtained by means of a 100×100 quadrilateral mesh (which leads to the same geometrical accuracy as the PGD).

Figure 3.5 depicts both solutions: the one related to the PGD on the left and the one associated with the application of the finite element on the right. The comparison of both solutions illustrates the performance of the PGD for computing the solution (the norm of the difference of both solutions is lower than 1 percent). It is important to recall that such solution defined in a space of dimension 27 cannot be computed by using a standard mesh based discretization technique. It was also verified that an

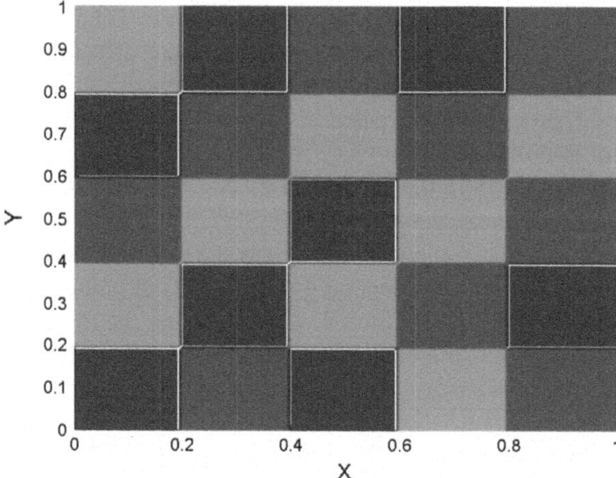

Fig. 3.4 Microstructure composed of 5×5 cells. Each color denotes a value of the thermal conductivity assumed isotropic and constant in each cell: *Blue*, *green* and *red* colors denote $k = 1$, $k = 1.5$ and $k = 2$ respectively

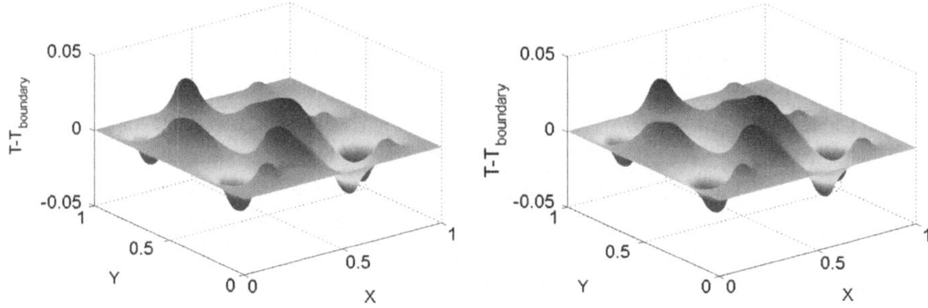

Fig. 3.5 $T(\mathbf{x}) - x$ computed by applying both the PGD (*left*), and the finite element method (*right*) for the microstructure depicted in Fig. 3.4 and for the boundary condition $T(\mathbf{x} \in \partial\Omega_{rve}) = x$

increasing number of terms N in the sum (3.40) makes the norm of the difference between the PGD and the FEM solutions decrease. Finally Fig. 3.6 depicts the functions $X_i(x)$, $Y_i(y)$, $\Xi^i_{11}(k_{1,1})$ and $\Xi^i_{55}(k_{5,5})$ involved in the decomposition (3.40) for $N = 83$.

3.3.2 Homogenization of Non-linear History Independent Models

In this section we focus on the homogenization of a simple nonlinear thermal model. The representative volume element $\Omega_{rve} = (0, 1) \times (0, 1)$ is composed of two materials of different conductivities k_1 and k_2. The zone $\omega = (x_l, x_r) \times (y_d, y_t) \subset$

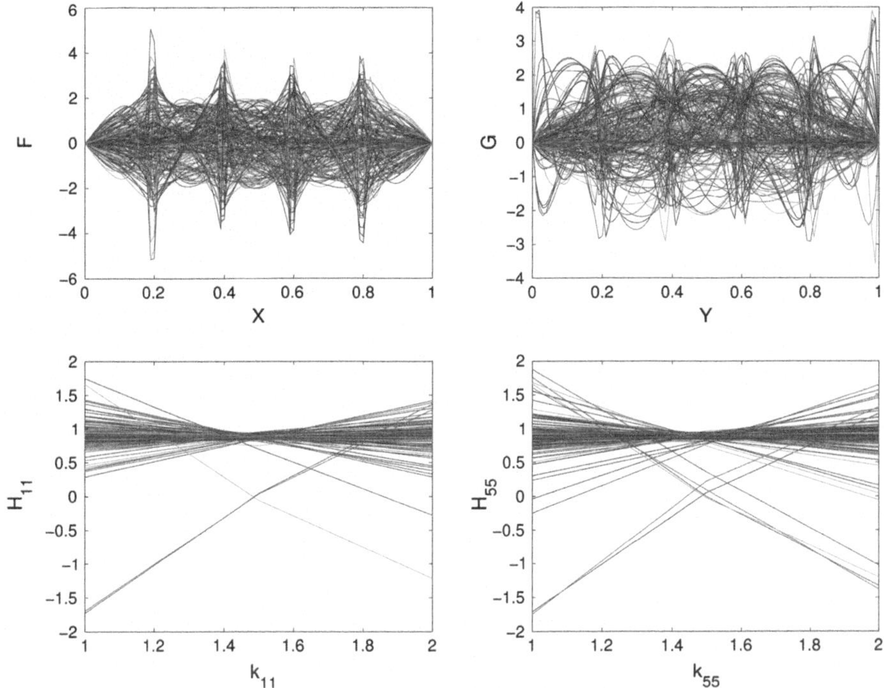

Fig. 3.6 Functions $X_i(x)$ (*top-left*), $Y_i(y)$ (*top-right*), $\Xi_{11}(k_{1,1})$ (*down-left*) and $\Xi_{55}(k_{5,5})$ (*down-right*)

Ω involves a material with conductivity k_2 whereas the material occupying the complementary region Ω-ω involves a material of conductivity k_1 (see Fig. 3.7).

The region of conductivity k_2 can be expressed by the following separate form:

$$\omega = \chi^x(x) \times \chi^y(y), \tag{3.42}$$

where both characteristic functions are defined by:

$$\chi^x(x) = \begin{cases} 1 & \text{if } x_l < x < x_r \\ 0 & \text{otherwise} \end{cases} \tag{3.43}$$

and

$$\chi^y(y) = \begin{cases} 1 & \text{if } y_d < x < y_t \\ 0 & \text{otherwise} \end{cases} \tag{3.44}$$

In the computations, we considered the following values for the geometrical parameters: $x_l = 0.4$, $x_r = 0.6$, $y_d = 0.4$ and $y_t = 0.6$. The microscopic conductivities are supposed to be linearly temperature dependent. As justified in Sect. 3.2.1, their values are obtained considering that the microscopic temperature equals the macroscopic

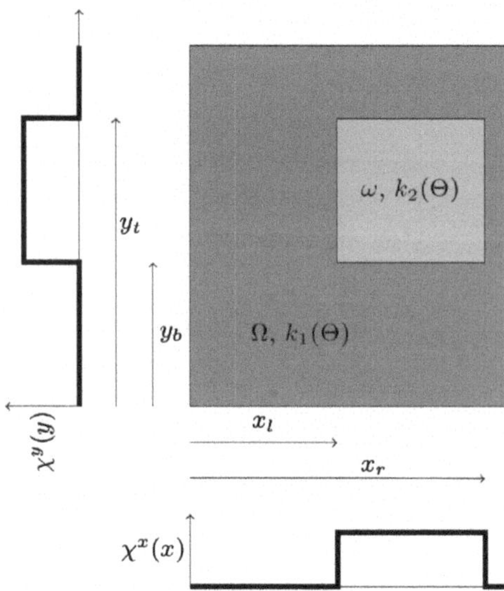

Fig. 3.7 Two phase microstructure

temperature Θ i.e., $k_1 = 10 + 9 \times \Theta$ and $k_2 = 90 + \Theta$, with $\Theta \in [0, 20]$. One can notice that this choice for the conductivities implies that when $\Theta = 10$ both conductivities take the same value and the microstructure becomes homogeneous.

With both conductivities defined and frozen during the microscopic step, the solution of the microscopic thermal model can be computed as previously for two different boundary conditions $T(\mathbf{x} \in \partial\Omega_{rve}) = x$ and $T(\mathbf{x} \in \partial\Omega_{rve}) = y$, leading to the homogenized conductivity tensor. If one proceeds in this manner, as soon as the macroscopic temperature Θ evolves in time, the homogenized conductivity tensor should be recomputed for the thermal properties related to the new macroscopic temperature. As discussed in Sect. 3.2.1 this procedure is very expensive from a computational point of view. An alternative lies in the solution of those microscopic models for any value of the macroscopic temperature, i.e. the calculation of $T^1(x, y, \Theta)$ and $T^2(x, y, \Theta)$ allowing us to solve two higher dimensional problems from which the temperature fields can be computed. Then, the homogenized conductivity tensor can be obtained for any macroscopic temperature.

In order to illustrate the feasibility of this approach, both solutions $T^1(x, y, \Theta)$ and $T^2(x, y, \Theta)$ are computed. Then, the resulting temperature field for a given macroscopic temperature $\Theta = \Theta^g$ (obtained by particularizing the general solution, $T^1(x, y, \Theta = \Theta^g)$ is compared with the one computed by applying the finite element method (with the thermal properties defined at temperature Θ^g). These results are depicted in Fig. 3.8 (left PGD, right FE) proving the ability of the multidimensional model to represent accurately the solution of any particular scenario, since the norm of the difference between the PGD and FE solutions is very small.

Finally Fig. 3.9 shows the most significant functions $X_i(x)$, $Y_i(y)$ and $\theta_i(\Theta)$ $(i = 1, \ldots, 100)$ related to the proper generalized decomposition. In this simulation,

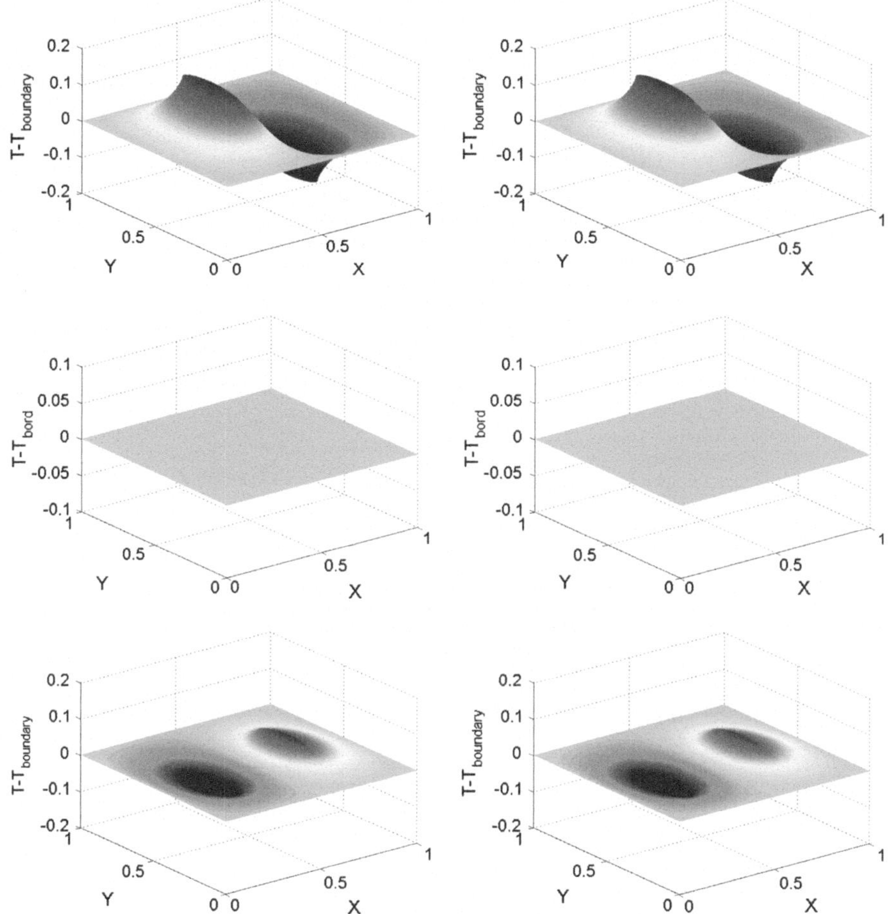

Fig. 3.8 $T(\mathbf{x}) - x$ computed by applying the PGD (*left*) and the FEM (*right*) for different macroscopic temperatures $\Theta = 0$ (*top*), $\Theta = 10$ (*middle*) and $\Theta = 20$ (*down*), when the thermal model in Ω_{rve} was solved by assuming $T(\mathbf{x} \in \partial\Omega_{rve}) = x$

1001 nodes were considered for the discretization of functions of x and y and 200 for those depending on the macroscopic temperature Θ.

From these results we can conclude on the ability of the PGD to capture the main features of the homogenized thermal model related to nonlinear behaviors. By solving a slightly higher dimensional model involving an extra-coordinate (the macroscopic temperature at the location of the RVE Ω_{rve}), the necessity of solving a microscopic model for each value of the macroscopic temperature can be avoided.

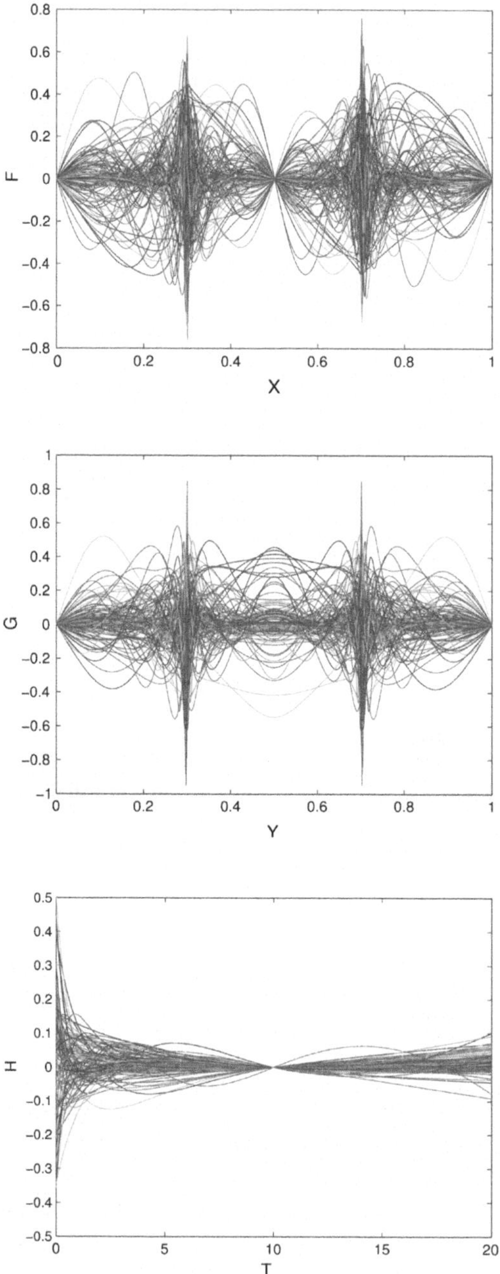

Fig. 3.9 Functions $X_i(x)$ (*top-left*), $Y_i(y)$ (*top-right*) and $\theta_i(\Theta)$ (*down*)

3.3.3 Homogenization of Non-linear History-Dependent Models

In this case we consider the scenario described in the previous example, but also assume that the thermal properties depend on the thermal history. Thus, a transient thermal model should be solved. Moreover, as the thermal properties at each time depend on the thermal history, a microscopic thermal model should be solved at each time step. In addition, the thermal properties in the RVE are assumed fixed (equal to $\Theta(\tau), 0 \leq \tau \leq t$), following the simplified model described in Sect. 3.2.4. To alleviate this solution one could try to solve the microscopic thermal model for any thermal history, from which the thermal properties in Ω_{rve} could be computed and frozen.

Following the description given in Sect. 3.2.4, the approximation of $\Theta(t)$ in the whole time interval was interpolated from a polynomial of degree 4 that can be expressed from 5 nodal values of Θ, denoted by $\Theta_i, i = 1, \dots M$ where M was set to $M = 5$ in our simulations.

Finally, the solutions for the two different boundary conditions are sought in the separated form $T^1(x, y, \Theta_1, \dots \Theta_M)$ and $T^2(x, y, \Theta_1, \dots \Theta_M)$ according to Eq. (3.39). In this case, two problems defined in a space involving x, y, t and the five temperatures $\Theta_i, i = 1, \dots, 5$ (8 dimensions) need to be solved in order to compute the homogenized conductivity tensor.

Functions depending on the space (x and y) where discretized by employing 101 nodes, whereas functions of time were integrated using also a coarse description involving 101 nodes. The length of the space and time interval were fixed arbitrarily to one. Finally the five extra-coordinates related to the intermediate macroscopic temperatures $\Theta_i, i = 1, \dots, 5$ were discretized by employing 10 nodes uniformly distributed in the interval of variation of those temperatures ([0, 10] here).

After computing both solutions $T^1(x, y, \Theta_1, \dots \Theta_M)$ and $T^2(x, y, \Theta_1, \dots \Theta_M)$ for any macroscopic thermal history described by the fourth order polynomial just defined, these solutions are particularized for a thermal history given by a linear thermal evolution from zero to 10 and then it decreases again linearly until reaching a null temperature at the final time. This problem, as soon as the thermal history is fixed, was also solved by mean of the finite element method. To validate the proposed approach we depict in Fig. 3.10 the solutions computed at the time $t = 0.25, t = 0.5$, $t = 0.75$ and $t = 1$ obtained by particularizing the solution $T^1(x, y, \Theta_1, \cdots \Theta_M)$ for the macroscopic thermal history just defined, i.e., $T^1(x, y, \Theta_1 = 0, \Theta_2 = 5, \Theta_3 = 10, \Theta_4 = 5, \Theta_5 = 0)$. The corresponding solutions, computed by applying the finite element solution and enforcing $T(\mathbf{x} \in \partial\Omega_{rve}, t) = x$ are depicted in Fig. 3.11. Both solutions agree in minute detail.

These results prove the ability of the proposed technique for solving nonlinear homogenization problems related to models involving internal variables implying a history dependence. Obviously a deeper analysis should be carried out for validating this approach in the case of real dependences of thermal parameters on the thermal history as well as for analyzing the sensibility of the results to the regularity in that thermal dependence.

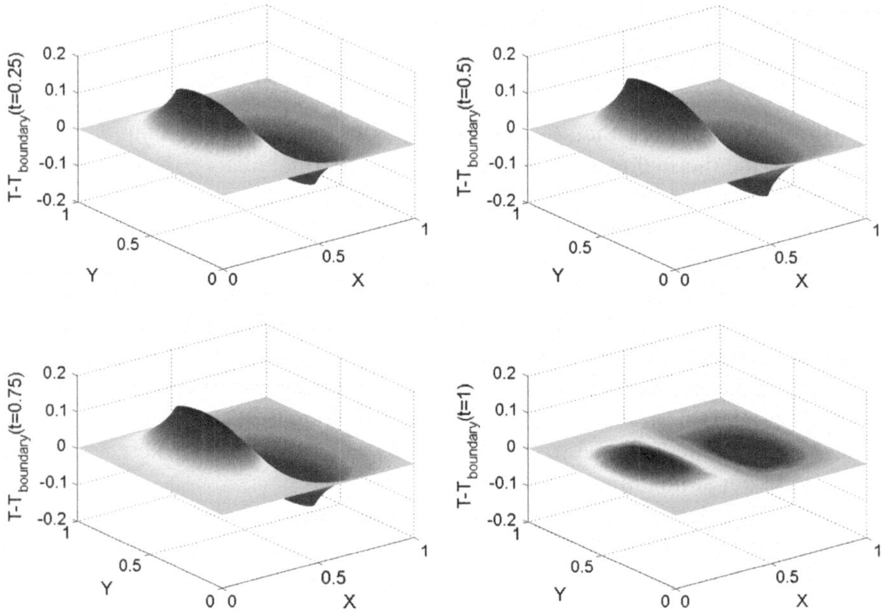

Fig. 3.10 $T(\mathbf{x}) - x$ computed by applying the PGD for $\Theta_1 = 0$, $\Theta_2 = 5$, $\Theta_3 = 10$, $\Theta_4 = 5$ and $\Theta_5 = 0$ at times $t = 0.25$ (*top-left*), $t = 0.5$ (*top-right*), $t = 0.75$ (*down-left*) and $t = 1$ (*down-right*) when the thermal model in Ω_{rve} was solved by assuming $T(\mathbf{x} \in \partial\Omega_{rve}) = x$

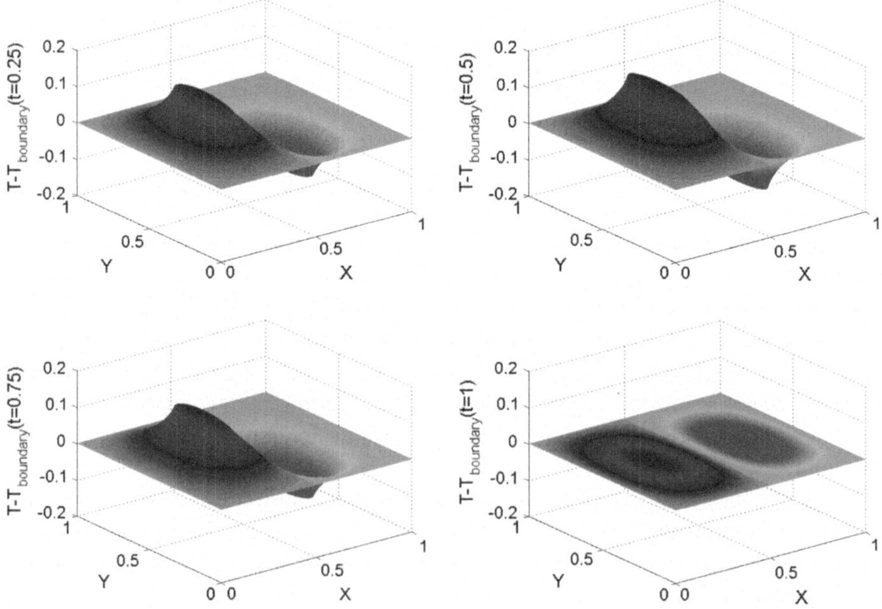

Fig. 3.11 $T(\mathbf{x}) - x$ computed by applying the FEM for $\Theta_1 = 0$, $\Theta_2 = 5$, $\Theta_3 = 10$, $\Theta_4 = 5$ and $\Theta_5 = 0$ at times $t = 0.25$ (*top-left*), $t = 0.5$ (*top-right*), $t = 0.75$ (*down-left*) and $t = 1$ (*down-right*) when the thermal model in Ω_{rve} was solved by assuming $T(\mathbf{x} \in \partial\Omega_{rve}) = x$

3.4 Conclusions

This work enlarges the domain of applicability of proper generalized decomposition to computational homogenization [32, 33]. In particular it proves that in the linear case it could be possible to define the homogenized thermal conductivity tensor for a generic microstructure composed of cells by assuming the conductivity of those cells as extra-coordinates. Even if the solution of the resulting multi-dimensional model could be expensive, it is important to recall that this solution should be carried out only once. Using a laptop and a handmade code in Matlab we solved models involving a 2D microstructure composed of 10×10 cells. This model was defined in a space involving $2 + 10 \times 10 = 102$ dimensions, and its solutions were performed without major difficulties. The computational resources available at present allows expecting new developments in this direction.

In the nonlinear case, we proposed a fully non-concurrent homogenization procedure by solving the microstructure for a generic macroscopic solicitation (macroscopic temperature or macroscopic gradient), even in the transient case.

All these developments were possible thanks to the introduction of a certain number of extra-coordinates in the models. The redoubtable curse of dimensionality related to multidimensional models was circumvented by employing separated representations, whose numerical complexity scales linearly with the dimension of the space in which the model is defined, instead of the exponential growing characteristic of mesh based representations.

References

1. J. Fish, Bridging the scales in nano engineering and science. J. Nanopart. Res. **8**(5), 577–594 (2006)
2. R. De Borst, Challenges in computational materials science: multiple scales, multi-physics and evolving discontinuities. Comput. Mater. Sci. **43**, 1–15 (2008)
3. C. McVeigh, W.K. Liu, Linking microstructure and properties through a predictive multiresolution continuum. Comput. Methods Appl. Mech. Eng. **197**(41–42), 3268–3290 (2008)
4. T. Zohdi, P. Wriggers, *An Introduction to Computational Micromechanics* (Springer, Berlin, 2005)
5. H.J. Bőhm, *A Short Introduction to Basic Aspects of Continuum Micromechanics, ISLB Report 208* (TU Wien, Vienna, 2009)
6. T.J.R. Hughes, Multiscale phenomena: green's functions, the dirichlet-to-neumann formulation, subgrid scale models, bubbles and the origins of stabilized methods. Comput. Methods Appl. Mech. Eng. **127**(1–4), 387–401 (1995)
7. J. Fish, Z. Yuan, Multiscale enrichment based on partition of unity. Int. J. Numer. Meth. Eng **62**(10), 1341–1359 (2005)
8. T. Strouboulis, L. Zhang, I. Babuška, P-version of the generalized fem using mesh-based handbooks with applications to multiscale problems. Int. J. Numer. Meth. Eng. **60**(10), 1639–1672 (2004)
9. F. Feyel, Multiscale fe2 elastoviscoplastic analysis of composite structures. Comput. Mater. Sci. **16**(1–4), 344–354 (1999)

10. M.G.D. Geers, V.G. Kouznetsova, W.A.M. Brekelmans, Multi-scale computational homogenization: trends and challenges. J. Comput. Appl. Math. (2009 in press). doi:10.1016/j.cam.2009.08.077
11. J.C. Michel, P. Suquet, Computational analysis of nonlinear composite structures using the nonuniform transformation field analysis. Comput. Methods Appl. Mech. Eng. **193**(48–51), 5477–5502 (2004)
12. I. Temizer, P. Wriggers, An adaptive method for homogenization in orthotropic nonlinear elasticity. Comput. Methods Appl. Mech. Eng. **196**(35–36), 3409–3423 (2007)
13. I. Temizer, T. Zohdi, A numerical method for homogenization in non-linear elasticity. Comput. Mech. **40**(2), 281–298 (2007)
14. J. Yvonnet, D. Gonzalez, Q.-C. He, Numerically explicit potentials for the homogenization of nonlinear elastic heterogeneous materials. Comput. Methods Appl. Mech. Eng. **198**(33–36), 2723–2737 (2009)
15. T. Kanit, S. Forest, I. Galliet, V. Mounoury, D. Jeulin, Determination of the size of the representative volume element for random composites: statistical and numerical approach. Int. J. Solids Struct. **40**(13–14), 3647–3679 (2003)
16. T. Kanit, F. N'Guyen, S. Forest, D. Jeulin, M. Reed, S. Singleton, Apparent and effective physical properties of heterogeneous materials: representativity of samples of two materials from food industry. Comput. Methods Appl. Mech. Eng. **195**(33–36), 3960–3982 (2006)
17. M. Ostoja-Starzewski, Material spatial randomness: from statistical to representative volume element. Probab. Eng. Mech. **21**(2), 112–132 (2006)
18. K. Sab, On the homogenization and the simulation of random materials. Eur. J. Mech. A. Solids **11**(5), 585–607 (1992)
19. M. Jiang, I. Jasiuk, M. Ostoja-Starzewski, Apparent thermal conductivity of periodic two-dimensional composites. Comput. Mater. Sci. **25**(3), 329–338 (2002)
20. I. Ozdemir, W.A.M. Brekelmans, M.G.D. Geers, Computational homogenization for heat conduction in heterogeneous solids. Int. J. Numer. Meth. Eng. **73**(2), 185–204 (2008)
21. P. Arbenz, G.H. van Lenthe, U. Mennel, R. Műller, M. Sala, A scalable multi-level preconditioner for matrix-free μ-finite element analysis of human bone structures. Int. J. Numer. Meth. Eng. **73**(7), 927–947 (2008)
22. T.G. Kolda, B.W. Bader, Tensor decompositions and applications. Technical Report (SAND2007-6702), SANDIA National Laboratories, Nov 2007
23. A. Ammar, F. Chinesta, P. Joyot, The nanometric and micrometric scales of the structure and mechanics of materials revisited: an introduction to the challenges of fully deterministic numerical descriptions. Int. J. Multiscale Comput. Eng. **6**(3), 191–213 (2008)
24. F. Chinesta, G. Chaidron, On the steady solution of linear advection problems in steady recirculating flows. J. NonNewton. Fluid Mech. **98**, 65–80 (2001)
25. G. Chaidron, F. Chinesta, On the steady solution of non-linear advection problems in steady recirculating flows. Comput. Methods Appl. Mech. Eng. **191**, 1159–1172 (2002)
26. F. Chinesta, G. Chaidron, A. Poitou, On the solution of the fokker-planck equations in steady recirculating flows involving short fiber suspensions. J. NonNewton. Fluid Mech. **113**(2–3), 97–125 (2003)
27. A. Ammar, B. Mokdad, F. Chinesta, R. Keunings, A new family of solvers for some classes of multidimensional partial differential equations encountered in kinetic theory modeling of complex fluids. part ii. J. NonNewton. Fluid Mech. **144**, 98–121 (2007)
28. S. Tokarzewski, I. Andrianov, Effective coefficients for real non-linear and fictitious linear temperature-dependent periodic composites. Int. J. Non-Linear Mech. **36**(1), 187–195 (2001)
29. J.J. Telega, S. Tokarzewski, A. Galka, Effective conductivity of nonlinear two-phase media: homogenization and two-point pade approximants. Acta Appl. Math. **61**(1), 295–315 (2000)
30. G. Laschet, Homogenization of the thermal properties of transpiration cooled multi-layer plates. Comput. Methods Appl. Mech. Eng. **191**(41–42), 4535–4554 (2002)
31. A. Alzina, E. Toussaint, A. Beakou, B. Skoczen, Multiscale modelling of thermal conductivity in composite materials for cryogenic structures. Compos. Struct. **74**(2), 175–185 (2006)

32. A. Mobasher-Amini, D. Dureisseix, P. Cartraud, Multi-scale domain decomposition method for large-scale structural analysis with a zooming technique: application to plate assembly. Int. J. Numer. Meth. Eng. **79**(4), 417–443 (2009)
33. P. Kanouté, D.P. Boso, J.L. Chaboche, B.A. Schrefler, Multiscale methods for composites: a review. Arch. Computat. Methods Eng. **16**(1), 31–75 (2009)

Chapter 4
Separated Representations of Coupled Models

In this chapter we explore some new possibilities based on separated space-time representations, that are closely inspired from some existing and well established strategies [1, 2]. We consider the coupling between global and many species kinetic local models and the issue related to the existence of different characteristic times of both the local and global model. This issue was also addressed in the context of proper generalized decompositions in [3].

4.1 Efficient Coupling of Global and Local Models

We are considering a simple parabolic model in a one-dimensional physical space Ω:

$$\frac{\partial u}{\partial t} - a\Delta u = f(x,t) \quad \text{in } \Omega \times (0, T_{\max}], \tag{4.1}$$

with the following initial and boundary conditions,

$$\begin{cases} u(x,0) = u^0 & x \in \Omega, \\ u(x,t) = u_g & (x,t) \in \partial\Omega \times (0, T_{\max}]. \end{cases} \tag{4.2}$$

We are assuming that the source term depends on the local value of r fields $C_i(x,t), i = 1, \ldots, r$:

$$f(x,t) = \sum_{i=1}^{i=r} \gamma_i \cdot C_i(x,t), \tag{4.3}$$

where the time evolution of the r fields $C_i(x,t)$ is governed by r coupled ordinary differential equations (the so-called kinetic model). For the sake of simplicity we consider the linear case, the nonlinear one reduces to a sequence of linear problems by applying an appropriate linearization strategy [4]. The system of linear ODEs is written at each point $x \in \Omega$:

F. Chinesta and E. Cueto, *PGD-Based Modeling of Materials, Structures and Processes*, ESAFORM Bookseries on Material Forming, DOI: 10.1007/978-3-319-06182-5_4,

$$\frac{dC_i(x,t)}{dt} = \sum_{j=1}^{j=r} \alpha_{ij}(x)\, C_j(x,t). \tag{4.4}$$

It is assumed that the kinetic coefficients α_{ij} evolve smoothly in Ω, since in practical applications these coefficients depend on the solution of the diffusion problem, $u(x,t)$. For the sake of simplicity and without loss of generality, α coefficients will be assumed later evolving linearly in x, but in the description that follows we assume those coefficients constant.

Now, we describe three possible procedures for solving Eqs. (4.1) and (4.4).

1. The simplest strategy consists in using a separated representation of the global problem solution (4.1) whereas the local problems are integrated in the whole time interval at each nodal position (or integration point). Obviously, this strategy implies the solution of r coupled ordinary differential equations at each node (or integration point). Moreover, the resulting fields $C_i(x,t)$, $r = 1, \ldots, r$, do not have a separated structure, and by this reason before injecting these fields into the global problem (4.1) we should separate them by invoking, for example, the singular value decomposition (SVD) leading to:

$$C_i(x,t) \approx \sum_{q=1}^{q=m} X_q^{C,i}(x) \cdot T_q^{C,i}(t). \tag{4.5}$$

   As soon as the source term has a separated structure, the procedure described in the previous chapters can be applied again for computing the new trial solution of the global problem that writes:

$$u(x,t) \approx \sum_{i=1}^{N} X_i^u(x) \cdot T_i^u(t). \tag{4.6}$$

   Thus, this coupling strategy requires the solution of many local problems (for all the coupled species at all nodal positions or integration points). Moreover, after these solutions (which we recall could be performed in parallel) a singular value decomposition must be applied in order to separate these solutions prior to injecting them in the PGD solver of the global problem (4.1).
2. The second coupling strategy lies in globalizing the solution of the local problems. Thus, we assume that the field related to each species can be written in a separated form:

$$C_i(x,t) \approx \sum_{q=1}^{q=m} X_q^{C,i}(x) \cdot T_q^{C,i}(t), \tag{4.7}$$

   and now, by applying the standard procedure to build up the reduced separated approximation, i.e. for constructing all the functions involved in (4.7). Thus, instead of solving the r coupled ODEs in Eq. (4.5) at each nodal position

(or integration point), we should solve only r higher dimensional coupled models defined in the physical space and time. Obviously, if the number of nodes (or integration points) is important (mainly when 3D physical spaces are considered) the present coupling strategy could offer significant CPU time savings.
This strategy allows computing directly a separated representation, and then, with respect to the previous one, the application of the SVD is avoided. However, if the number of species is high, the computational efforts can become important, because the space-time separated solver must be applied to each species.

3. The third alternative, that in our opinion is the more appealing one for solving models involving many species, as many as one wants, implies the definition of a new variable $C(x, t, c)$, that as we can notice contains an extra coordinate c, with discrete nature, and that takes integer values: $c = 1, \ldots, r$, in such manner that $C(x, t, i) \equiv C_i(x, t)$, $i = 1, \ldots, r$. Thus, we have increased the dimensionality of the problem, but now, only a single problem should be solved, instead one for each species as was the case when using the previous strategy. This increase of the model dimensionality is not dramatic because the separated representation allows circumventing the curse of dimensionality, allowing for fast and accurate solutions of highly multidimensional models. Now, the issue is the derivation of the governing equation for this new variable $C(x, t, c)$ and the separated representation constructor able to define the approximation:

$$C(x, t, c) \approx \sum_{q=1}^{q=S} X_q^C(x) \cdot T_q^C(t) \cdot A_q(c). \tag{4.8}$$

Since this strategy will be retained in our simulations we will focus on its associated computational aspects in the next section.

4.2 Fully Globalized Local Models

The third strategy just cited implies the solution of a single multidimensional model involving the field $C(x, t, c)$. This original introduction deserves some additional comments. The first one concerns the discrete nature of the kinetic equations

$$\frac{dC_i(x, t)}{dt} = \sum_{j=1}^{j=r} \alpha_{ij}(x) \cdot C_j(x, t). \tag{4.9}$$

Now, by introducing $C(x, t, c)$, such that $C(x, t, i) \equiv C_i(x, t)$, the kinetic equations could be written as:

$$\frac{dC}{dt} = \mathscr{L}_c(C), \tag{4.10}$$

where $\mathscr{L}_c$ is an operator in the c-coordinate.

If for one second we try to discretize Eq. (4.10) by finite differences, we could write at each node (x_k, t_p, i):

$$\frac{C(x_k, t_p, i) - C(x_k, t_{p-1}, i)}{\Delta t} = \mathscr{L}_c(C)|_i \,, \tag{4.11}$$

where

$$\mathscr{L}_c(C)|_i = \sum_{j=1}^{j=r} \alpha_{ij} \cdot C(x_k, t_p, j), \tag{4.12}$$

represents the discrete form of the c-operator.

We now come back to the separated representation constructor for defining the approximation:

$$C(x,t,c) \approx \sum_{q=1}^{q=S} X_q^C(x) \cdot T_q^C(t) \cdot A_q(c). \tag{4.13}$$

To define such approximation one should repeat the PGD constructor procedure. As the operator here involved is less standard and concerns a discrete coordinate we summarize the main steps.

Asume that the first n iterations allows computing the first n sums of Eq. (4.13)

$$C(x,t,c) \approx \sum_{q=1}^{q=n} X_q^C(x) \cdot T_q^C(t) \cdot A_q(c), \tag{4.14}$$

and now, we look for the enrichment $R(x) \cdot S(t) \cdot W(c)$, such that

$$\begin{aligned} C(x,t,c) &\approx \sum_{q=1}^{q=n} X_q^C(x) \cdot T_q^C(t) \cdot A_q(c) + R(x) \cdot S(t) \cdot W(c) \\ &= C^n(x,t,c) + R(x) \cdot S(t) \cdot W(c), \end{aligned} \tag{4.15}$$

satisfies

$$\int_{\Omega} \int_0^{T_{\max}} \int_0^r C^*(x,t,c) \cdot \left(\frac{dC}{dt} - \mathscr{L}_c(C) \right) dc \; dt \; dx = 0. \tag{4.16}$$

Obviously, due to the discrete character of the third coordinate, an integration quadrature consisting of r points, $c_1 = 1, \ldots, c_r = r$ will be considered later.

To compute the three enrichment functions we consider again an alternating directions strategy, that proceeds in three steps (that are repeated until reaching convergence):

1. Assuming functions $S(t)$ and $W(c)$ known, the trial function $C^*(x, t, c)$ writes $R^*(x) \cdot S(t) \cdot W(c)$. Thus the weak form (4.16) reads:

$$\int_{\Omega} \int_0^{T_{\max}} \int_0^r R^* \cdot S \cdot W \cdot \left(R \cdot S' \cdot W - R \cdot S \cdot \mathscr{L}_c(W) \right) \, dc \, dt \, dx$$

$$= - \int_{\Omega} \int_0^{T_{\max}} \int_0^r R^* \cdot S \cdot W \cdot \sum_{q=1}^{q=n} \left(X_q^C \cdot (T_q^C)' \cdot A_q \right) \, dc \, dt \, dx$$

$$+ \int_{\Omega} \int_0^{T_{\max}} \int_0^r R^* \cdot S \cdot W \cdot \sum_{q=1}^{q=n} \left(X_q^C \cdot T_q^C \cdot \mathscr{L}_c(A_q) \right) \, dc \, dt \, dx, \quad (4.17)$$

where $S' = \frac{dS}{dt}$ and $(T_q^C)' = \frac{dT_q^C}{dt}$.
Time integrals and the ones involving the c-coordinate can now be performed. The ones involving time are easily performed. The ones involving the c-coordinate are written:

$$\int_0^r W \cdot W \, dc = \sum_{i=1}^{i=r} W(c_i)^2, \quad (4.18)$$

where as just mentioned $c_i = i, \forall i$,

$$\int_0^r W \cdot \mathscr{L}_c(W) \, dc = \sum_{i=1}^{i=r} \left(W(c_i) \cdot \sum_{j=1}^{j=r} \alpha_{ij} W(c_j) \right), \quad (4.19)$$

and similar expressions can be derived for the integrals involved in the right-hand member.
Thus, finally we get:

$$\xi^x \int_{\Omega_x} R^* \cdot R \, dx = \int_{\Omega_x} R^* F^x(x) \, dx, \quad (4.20)$$

where the coefficient ξ^x contains all the integrals in the time and c-coordinates related to the left hand member of Eq. (4.17) and $F^x(x)$ all the integrals appearing in the right hand member. The strong form related to Eq. (4.20) is written:

$$\xi^x R(x) = F^x(x), \quad (4.21)$$

whose algebraic nature derives from the fact that kinetic model is local and then it does not involve space derivatives.

2. Assuming functions $R(x)$ and $W(c)$ known, the trial function $C^*(x,t,c)$ is written $R(x) \cdot S^*(t) \cdot W(c)$. Thus the weak form (4.16) reads:

$$\int_{\Omega}\int_0^{T_{\max}}\int_0^{r} R \cdot S^* \cdot W \cdot \left(R \cdot S' \cdot W - R \cdot S \cdot \mathscr{L}_c(W)\right)\, dc\, dt\, dx$$
$$= - \int_{\Omega}\int_0^{T_{\max}}\int_0^{r} R \cdot S^* \cdot W \cdot \sum_{q=1}^{q=n} \left(X_q^C \cdot (T_q^C)' \cdot A_q\right)\, dc\, dt\, dx$$
$$+ \int_{\Omega}\int_0^{T_{\max}}\int_0^{r} R \cdot S^* \cdot W \cdot \sum_{q=1}^{q=n} \left(X_q^C \cdot T_q^C \cdot \mathscr{L}_c(A_q)\right)\, dc\, dt\, dx. \quad (4.22)$$

Integrals defined in the physical space Ω must now be computed, but this task does not involve additional difficulties.
Finally it results:

$$\int_0^{T_{\max}} S^* \cdot \left(\xi^t S + \upsilon^t \frac{dS}{dt}\right) dt = \int_0^{T_{\max}} S^* F^t(t)\, dt, \quad (4.23)$$

where coefficients ξ^t and υ^t contain all the integrals in the space and the c-coordinate related to the left-hand member of Eq. (4.22) and $F^t(t)$ the associated integrals appearing in the right-hand member. The strong form related to Eq. (4.23) is written:

$$\upsilon^t \frac{dS}{dt} + \xi^t S(t) = F^t(t), \quad (4.24)$$

whose first-order differential nature derives from the form of kinetic models.

3. Assuming functions $R(x)$ and $S(t)$ known, the trial function $C^*(x,t,c)$ writes $R(x) \cdot S(t) \cdot W^*(c)$. Thus the weak form (4.16) reads:

$$\int_{\Omega}\int_0^{T_{\max}}\int_0^{r} R \cdot S \cdot W^* \cdot \left(R \cdot S' \cdot W - R \cdot S \cdot \mathscr{L}_c(W)\right)\, dc\, dt\, dx$$
$$= - \int_{\Omega}\int_0^{T_{\max}}\int_0^{r} R \cdot S \cdot W^* \cdot \sum_{q=1}^{q=n} \left(X_q^C \cdot (T_q^C)' \cdot A_q\right)\, dc\, dt\, dx$$
$$+ \int_{\Omega}\int_0^{T_{\max}}\int_0^{r} R \cdot S \cdot W^* \cdot \sum_{q=1}^{q=n} \left(X_q^C \cdot T_q^C \cdot \mathscr{L}_c(A_q)\right)\, dc\, dt\, dx. \quad (4.25)$$

After performing integration in space and time, we get that:

$$\int_0^r W^* \cdot \left(\xi^c\, W - \upsilon^c\, \mathcal{L}_c(W)\right)\, dc = \int_0^r W^* F^c(c)\, dc, \tag{4.26}$$

where coefficients ξ^c and υ^c contain all the integrals in the space and time related to the left hand member of Eq. (4.25) and $F^c(c)$ the associated integrals appearing in the right hand member. The strong form related to Eq. (4.26) writes:

$$-\,\upsilon^c\, \mathcal{L}_c(W) + \xi^c W(c) = F^c(c), \tag{4.27}$$

that results in the algebraic system:

$$-\,\upsilon^c \sum_{j=1}^{j=r} \alpha_{ij}\, W(c_j) + \xi^c W(c_i) = F^c(c_i), \quad i = 1, \ldots, r. \tag{4.28}$$

4.3 Heterogeneous Time Integration

In previous chapters we illustrated the construction of a separated representation of standard global models, as the one defined in Eq. (4.1). In the previous section we illustrated some original possibilities for globalizing local models consisting in kinetic equations. Now, the separated representation of the different kinetic fields C_i can be extracted from the general solution:

$$C(x,t,c) \approx \sum_{q=1}^{q=S} X_q^C(x) \cdot T_q^C(t) \cdot A_q(c), \tag{4.29}$$

by writing:

$$C_i(x,t) = C(x,t,c_i) = C(x,t,i) \approx \sum_{q=1}^{q=S} X_q^C(x) \cdot T_q^C(t) \cdot A_q(i). \tag{4.30}$$

Thus, we obtain a fast separated representation of the source term in Eq. (4.1) allowing us to solve this equation within the PGD framework. However, it remains a detail that deserves some comments. The characteristic time related to local and global problems could differ in some orders of magnitude. In that case functions T_q^C are defined using a time step δt much lower than the one employed for defining functions T_i^u appearing in the separated representation of the global problem solution

$$u(x,t) \approx \sum_{i=1}^{N} X_i^u(x) \cdot T_i^u(t). \tag{4.31}$$

Thus, the issue is knowing $C_i(x,t)$ with a time resolution of δt, which is the simplest consistent transfer to the solution of the global problem that is performed with a time resolution of Δt? It is easy to prove that the simplest choice consistent with a first-order discontinuous Galerkin time integration of the ODE involved in the computation of functions $T_i^u(t)$ (that ensures the conservation of the integrated variable) consists of defining $\overline{C}_i(x,t)$ with a resolution of Δt, such that

$$\overline{C}_i(x, t = n \times \Delta t) = \frac{1}{\Delta t} \int_{(n-1)\times\Delta t}^{n\times\Delta t} C_i(x,t)\, dt. \tag{4.32}$$

In any case, and contrary to the experiences in using standard incremental strategies, the use of the proper generalized decomposition reduced significantly the impact that heterogeneous time integrations have in the total amount of CPU time. This is due to the fact that the time step only affects to the solution of the one dimensional equations solved for computing functions T_i^u and T_q^C, and this step is much more faster than the one needed for computing the space functions in the global problem $X_i^u(\mathbf{x})$ that requires the solution of a steady state 2D or 3D problem. Thus, when the differences of the characteristic times of both models is not too large we could perfectly consider the lowest characteristic time step in the integration of both models.

4.4 Numerical Example

We illustrate the full globalization of local kinetic models thoroughly described in the previous section, because the solution of global models and their coupling do not involves major difficulties, and both aspects were reported in some of our previous works [4, 5].

To enforce a spatial dependence of the kinetic model solution we consider a kinetic model in which the kinetic coefficients evolve linearly in space. For the sake of simplicity we also consider a one-dimensional physical space and ten species (i.e. $r = 10$).

The associated kinetic equations is written:

$$\frac{dC_i(x,t)}{dt} = \sum_{j=1}^{j=r} \alpha_{ij}(x)\, C_j(x,t), \tag{4.33}$$

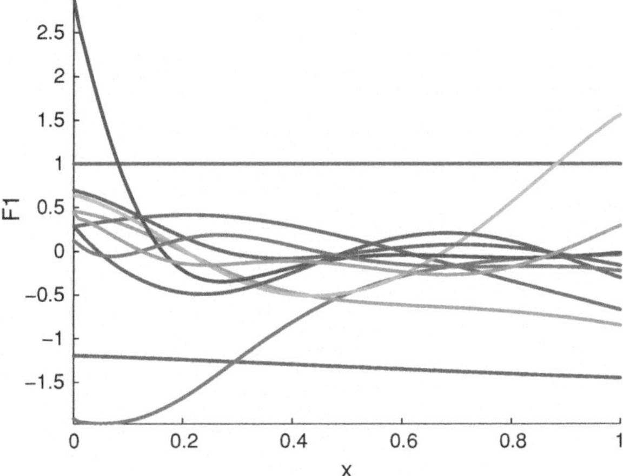

Fig. 4.1 Separated representation: spatial functions

where

$$\alpha_{ij}(x) = \alpha'_{ij} + x \cdot \alpha''_{ij}. \tag{4.34}$$

Coefficients α'_{ij} and α''_{ij} were defined by

$$\alpha'_{ij} = \frac{1}{r^2} i \times j - 0.5, \tag{4.35}$$

$$\alpha''_{ij} = \frac{1}{r^2}(r + 1 - i) \times (r + 1 - j) - 0.5. \tag{4.36}$$

The procedure previously described was applied and a separated representation of the field $C(x, t, c)$ was obtained after 11 iterations, i.e. the solution contains 11 sums:

$$C(x, t, c) \approx \sum_{q=1}^{q=11} X_q^C(x) \cdot T_q^C(t) \cdot A_q(c). \tag{4.37}$$

Figures 4.1, 4.2 and 4.3 depict the computed functions $X_q^C(x)$, $T_q^C(x)$ and $A_q^C(x)$, that were denoted by $F1$, $F2$ and $F3$ respectively. Obviously, in Fig. 4.3 only the values of the different curves for $c = 1, 2, \ldots, 10$ make sense.

Figures 4.4 and 4.5 depict the space-time distribution of species 1–5 and 6–10 respectively.

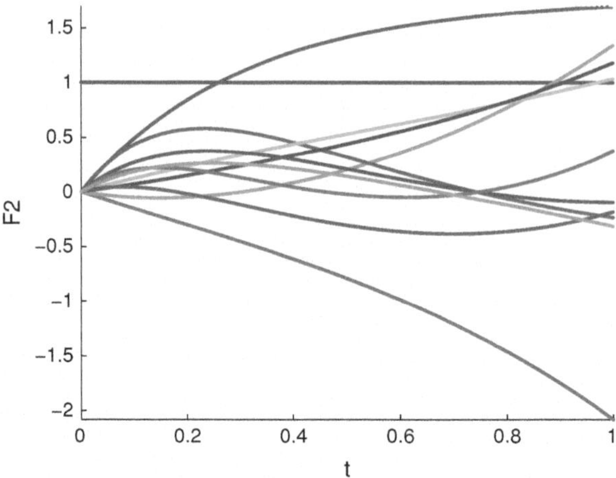

Fig. 4.2 Separated representation: time functions

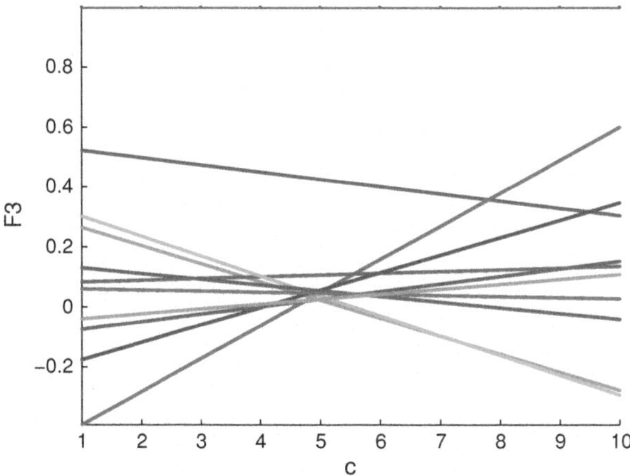

Fig. 4.3 Separated representation: functions defined on the species coordinate

4.5 Discussion

Because neither the PGD-based solver nor the incremental one are computationally optimized, we do not address the computing time involved in the simulation. However, we could quantify the advantage in using the PGD based solver by evaluating the number of operations involved in both solutions. We are assuming the simplest numerical choices: explicit incremental integration, …

We recall the notation previously introduced: N is the number of functional products involved in the separated approximation of field $C(x, t, c)$, Q being the number of iterations of the alternating direction fixed point algorithm needed for computing

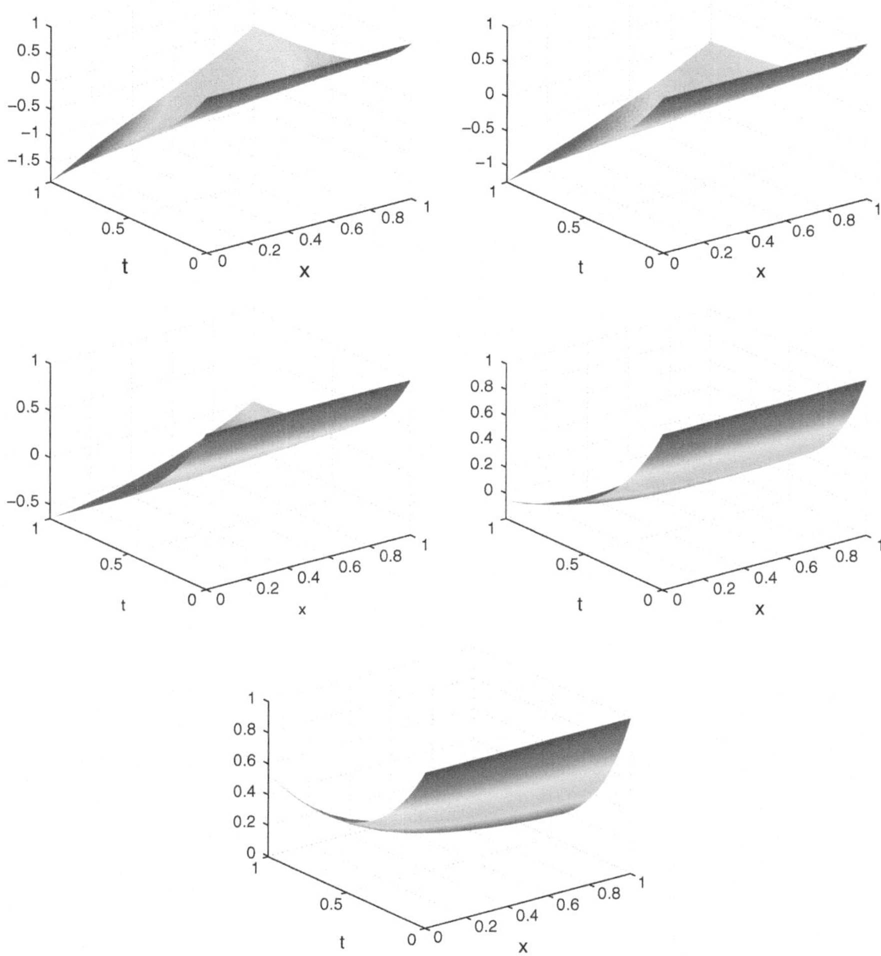

Fig. 4.4 Space-time distribution of species 1–5

each one of the functional products. We consider N_n nodes in which the dynamical system must be integrated, P being the number of time steps and r the number of species involved by the kinetic model.

When a standard incremental technique is used (without any kind of computational improvement) the number of operations scales with $N_n \times P \times r$, whereas when using the PGD based solver the number of operations scales with $N \times Q \times (N_n + r^2 + P)$.

When we consider a simple problem characterized by $N_n = 1000$, $P = 1000$ and $r = 10$, the PGD based solver requires of about $N \times Q \approx 1000$ iterations. Thus, the number of operations involved by the incremental technique is of order 10^7 whereas the one related to the PGD based solver requires about 10^6. In this case there is not

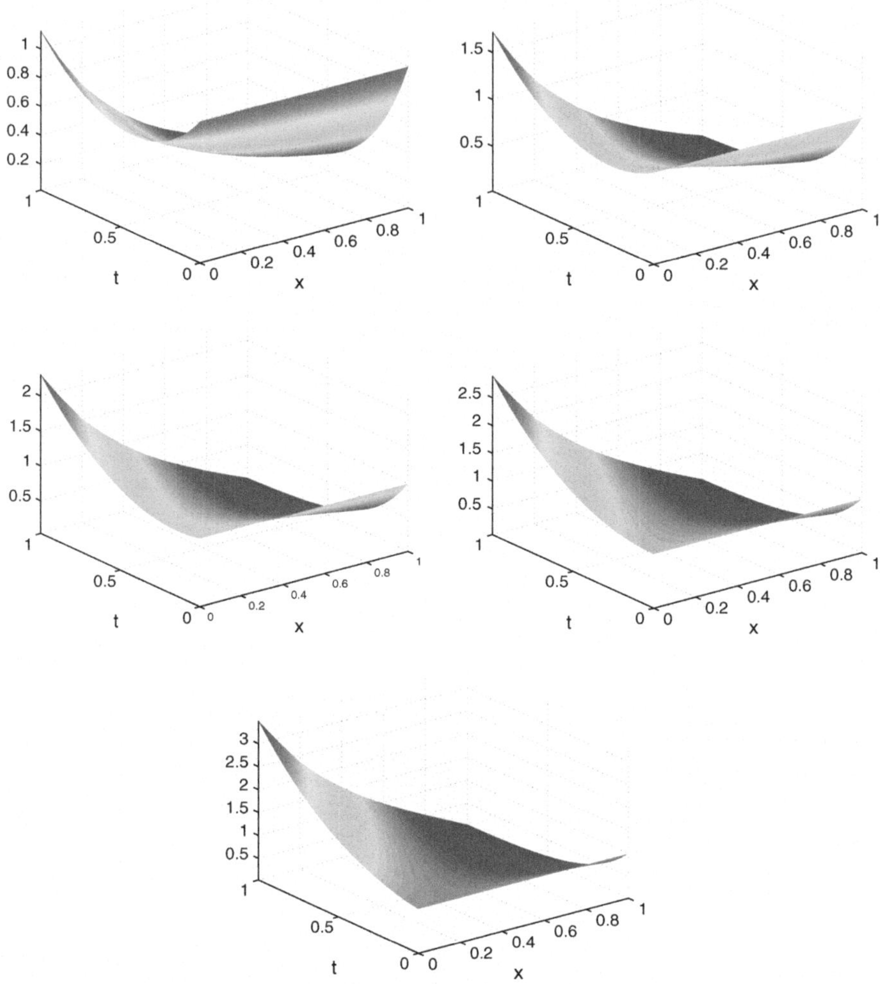

Fig. 4.5 Space-time distribution of species 6–10

a clear advantage in using the PGD based solver. Now, if $N_n = 10^6$, $P = 10^6$ and $r = 10$, the incremental solver needs 10^{13} operations instead of the 10^9 involved in the PGD based solver. In this case the CPU time savings is of four orders of magnitude, and this ratio can be increased by assuming more complex simulation scenarios involving more nodes, time steps or species.

References

1. A. Ammar, B. Mokdad, F. Chinesta, R. Keunings. A new family of solvers for some classes of multidimensional partial differential equations encountered in kinetic theory modeling of complex fluids. Part II: Transient simulation using space-time separated representation. J. Non-Newton. Fluid. Mech. **144**, 98–121 (2007)
2. P. Ladeveze, *Nonlinear Computational Structural Mechanics* (Springer, New York, 1999)
3. D. Néron, P. Ladevèze, Proper generalized decomposition for multiscale and multiphysics problems. Arch. Comput. Meth. Eng. **17**(4), 351–372 (2010)
4. A. Ammar, E. Pruliere, F. Chinesta, M. Laso, Reduced numerical modeling of flows involving liquid-crystalline polymeres. J. NonNewton. Fluid. Mech. **160**, 140–156 (2009)
5. E. Pruliere, A. Ammar, N. El Kissi, F. Chinesta, Recirculating flows involving short fiber suspensions: numerical difficulties and efficient advanced micro-macro solvers. Arch. Comput. Meth. Eng. State. Art. Rev. **16**, 1–30 (2009)

Chapter 5
Parametric Models in Evolving Domains

This chapter addresses the recurrent issue related to the existence of reduced bases related to the solution of parametric models defined in evolving domains. Here we address the case of decoupled kinematics, i.e., models whose solution does not affect the domain in which they are defined. The chosen framework considers an updated Lagrangian description of the kinematics, solved by using natural neighbor Galerkin methods within a non-incremental space-time framework that can be generalized for addressing parametric models. Examples are included showing the performance and potentialities of the proposed methodology.

5.1 On the Difficulty of Simulating Evolving Domains

Evolving domains introduce many numerical difficulties. Firstly, when using a fixed mesh the domain evolution must be captured by using an appropriate technique (e.g., VOF, level sets, or any similar technique). The resulting advection terms must be efficiently stabilized by using, in turn, an adequate technique (SUPG, DG, ...). Numerous works have addressed such questions during the last decades see, for instance, [1] and references therein.

Another possibility consists in tracking the domain whose geometry evolves with the material velocity, in an (updated) Lagrangian approach. This approach simplifies the treatment of advection terms that now result in a simple material derivative. The main drawback is, however, that meshes become rapidly too distorted implying the need for frequent remeshing and the associated field projection between old and new meshes. A particularly elegant analysis of the difficulties associated with this approach to the problem can be found in [2]. Intermediate procedures have been proposed in the framework of ALE methods alleviating partially the issues of fixed and moving meshes [1]. However, the determination of the optimal velocity of the mesh is a tricky problem.

F. Chinesta and E. Cueto, *PGD-Based Modeling of Materials, Structures and Processes*, ESAFORM Bookseries on Material Forming, DOI: 10.1007/978-3-319-06182-5_5,

Some years ago new discretization techniques were proposed whose accuracy proved to be independent of the nodal distribution used to approximate the different fields involved in the models. These techniques were called meshless or meshfree methods, even if some of them employ a background mesh to construct the functional approximation or even to perform numerical integration. This designation is justified by the fact that the approximation accuracy does not depend on the relative position of the nodes. As a result, remeshing can be avoided even in the case of large distortions of the background mesh.

Despite the chosen framework for the description of the kinematics, it is well known in the reduced order modeling community, that the determination of an efficient set of reduced basis for problems defined in an evolving domain is a difficult task. This is caused undoubtedly because the deformation of the domain very much complicates the concept of *snapshot*, which is crucial in understanding the overall behavior of the system, and in determining the set of reduced basis itself [3].

In this work, among many different possibilities, we have chosen the Natural Element Method (NEM), widely described in [4] and references therein, to approximate the kinematics in an updated Lagrangian framework. NEM overcomes FEM remeshing needs by employing natural neighbor approximation instead of piecewise polynomials to construct shape functions in a Galerkin setting. Thus, within the NEM framework one can proceed with the original cloud of nodes moving according to material velocity during the whole simulation, even in the case of very large geometrical transformations. This is not a crucial choice in the development that follows (many other meshless methods exist that allow for an updated Lagrangian description of kinematics), albeit NEM presents some very interesting characteristics that will be analyzed below.

In what follows we consider a model defined in a domain that at time $t = 0$ occupies the region $\Omega^0 \subset \mathbb{R}^3$. The different fields in this domain are approximated from a cloud of N_n nodes located at positions $\tilde{\mathbf{x}}_i^0, i = 1, \ldots, N_n$. The material domain evolves in time, $\Omega(t)$ representing its configuration at time t. We assume that this evolution is defined by a given, decoupled, velocity field $\mathbf{v}(\mathbf{x} \in \Omega(t), t \in \mathscr{I} \subset \mathbb{R}_+)$. Nodes move with the material velocity, and because the meshless behavior of the NEM approximation, all the fields are approximated in the updated domain $\Omega(t)$ by using the original cloud of nodes. No addition or deletion of nodes is considered, even if it is perfectly possible in a NEM framework. At time t nodal positions will be denoted by $\tilde{\mathbf{x}}_i^t, i = 1, \ldots, N_n$.

Hereafter we assume, without loss of generality, that the model, defined in the evolving domain $\Omega(t)$, involves the unknown field $u(\mathbf{x} \in \Omega(t), t \in \mathscr{I})$. We focus on the possibility of determining a reduced basis approximation for such a field in the context of the proper generalized decomposition (PGD) framework. In this work a strategy able to compute transient solutions in evolving domains is proposed. This strategy falls within a non-incremental framework originally proposed in a different context by Ladeveze [5]. Moreover, it will be shown how efficiently parametric models defined in evolving domains can be solved. Here, the model parameter, say the thermal conductivity k of the thermal model here addressed, could be introduced as an extra-coordinate in the model, and then a multidimensional representation of the unknown field $u(\mathbf{x} \in \Omega(t), t \in \mathscr{I}, k \in \Im)$ will be found.

5.2 Separated Representation of Models Defined in Evolving Domains

In order to show how the just explained strategy can be extended to problems defined in evolving domains, we come back to a non-parametric problem, for the sake of simplicity in the exposition (the parametric case will be addressed later) and consider the advection-diffusion equation defined in a domain $\Omega(t)$ that evolves with a prescribed velocity field $\mathbf{v}(\mathbf{x}, t)$, $\mathbf{x} \in \Omega(t)$ and $t \in \mathscr{I}$. Without loss of generality we assume homogeneous initial and boundary conditions. The issue related to the enforcement of non homogeneous boundary conditions was comprehensively addressed in [6].

The weak form of the problem, in this case, reads: *find* $u(\mathbf{x}, t)$ *such that*

$$\int_{\mathscr{I}} \int_{\Omega(t)} u^* \cdot \left(\frac{Du}{Dt} - k \cdot \Delta u - s \right) d\mathbf{x} \cdot dt = 0, \tag{5.1}$$

holds for every test function u^* defined in an appropriate Hilbert space. The source term is considered depending on the space and time coordinates, i.e. $s(\mathbf{x}, t)$.

By integrating by parts, the weak form is written:

$$\int_{\mathscr{I}} \int_{\Omega(t)} \left(u^* \cdot \frac{Du}{Dt} + k \cdot \nabla u^* \cdot \nabla u - u^* \cdot s \right) d\mathbf{x} \cdot dt = 0. \tag{5.2}$$

The approximation of the field $u(\mathbf{x}, t)$ is constructed from its nodal values $u_i^t \equiv u(\tilde{\mathbf{x}}_i^t, t)$ by utilizing a natural neighbor (NN) interpolation:

$$u(\mathbf{x} \in \Omega(t), t) \approx \sum_{i=1}^{i=N_n} N_i^t(\mathbf{x}) \cdot u_i^t = \mathbf{N}^t \cdot \mathbf{U}^t, \tag{5.3}$$

where the upper-index t associated to the shape functions N_i^t indicates that these shape functions where defined from the nodal positions $\tilde{\mathbf{x}}_i^t$ in $\Omega(t)$.

As mentioned before, although it is by no means the only possible choice, NN interpolation has remarkable properties that make their use in this context very convenient [4]. Undoubtedly, one of them is the Kroenecker delta property that states:

$$N_i^t(\tilde{\mathbf{x}}_j^t) = \delta_{ij}, \tag{5.4}$$

that, together with the exact interpolation on boundaries, makes it possible to easily enforce Dirichlet boundary conditions.

In what follows we analyze separately the different terms in Eq. (5.2).

5.2.1 Diffusive Term

We consider the diffusive term in Eq. (5.2):

$$\mathscr{D} = \int_{\mathscr{I}} \int_{\Omega(t)} k \cdot \nabla u^* \cdot \nabla u \, d\mathbf{x} \cdot dt. \tag{5.5}$$

If we define a matrix $\mathbf{B}^t$ containing the derivatives of the shape functions N_i^t:

$$\mathbf{B}^t = \begin{pmatrix} \frac{dN_1^t}{dx} & \frac{dN_2^t}{dx} & \cdots & \frac{dN_n^t}{dx} \\ \frac{dN_1^t}{dy} & \frac{dN_2^t}{dy} & \cdots & \frac{dN_n^t}{dy} \\ \frac{dN_1^t}{dz} & \frac{dN_2^t}{dz} & \cdots & \frac{dN_n^t}{dz} \end{pmatrix}. \tag{5.6}$$

The diffusive term can be written as:

$$\begin{aligned} \mathscr{D} &= \int_{\mathscr{I}} \int_{\Omega(t)} k \cdot \mathbf{U}^{*^T} \cdot \mathbf{B}^{t^T} \cdot \mathbf{B}^t \cdot \mathbf{U}^t \, d\mathbf{x} \cdot dt \\ &= \int_{\mathscr{I}} k \cdot \mathbf{U}^{*^T} \cdot \left(\int_{\Omega(t)} \mathbf{B}^{t^T} \cdot \mathbf{B}^t \, d\mathbf{x} \right) \cdot \mathbf{U}^t \, dt \\ &= \int_{\mathscr{I}} k \cdot \mathbf{U}^{*^T} \cdot \mathbf{G}(t) \cdot \mathbf{U}^t \, dt. \end{aligned} \tag{5.7}$$

Because $\Omega(t)$ is known $\forall t$, we can evaluate the integral

$$\mathbf{G}_k = \int_{\Omega(t_k)} \mathbf{B}^{t_k^T} \cdot \mathbf{B}^{t_k} \, d\mathbf{x}, \tag{5.8}$$

at different times t_k, $k = 1, \ldots, Q$.

Following the spirit of the proper orthogonal decompositions, from these integrals we could define a matrix $\mathbf{G}$

$$\mathbf{G} = \left(\mathbf{G}_1 \; \mathbf{G}_2 \; \cdots \; \mathbf{G}_Q \right), \tag{5.9}$$

that, after applying a singular value decomposition (SVD), gives

$$\mathbf{G}(t) = \int_{\Omega(t)} \mathbf{B}^{t^T} \cdot \mathbf{B}^t \, d\mathbf{x} \approx \sum_{j=1}^{j=m_1} F_j^d(t) \cdot \mathbf{E}_j^d, \tag{5.10}$$

with $m_1 < Q$ and $m_1 < N_n$.

Thus, the diffusive term can be advantageously written as:

$$\mathscr{D} = \int_{\mathscr{I}} \int_{\Omega(t)} k \cdot \mathbf{U}^{*^T} \cdot \mathbf{B}^{t^T} \cdot \mathbf{B}^t \cdot \mathbf{U}^t \, d\mathbf{x} \cdot dt = \int_{\mathscr{I}} k \cdot \mathbf{U}^{*^T} \cdot \left(\sum_{j=1}^{j=m_1} F_j^d(t) \cdot \mathbf{E}_j^d \right) \cdot \mathbf{U}^t \, dt. \tag{5.11}$$

5.2.2 Advective Term

We consider now the term involving time derivatives:

$$\mathscr{A} = \int_{\mathscr{I}} \int_{\Omega(t)} u^* \cdot \frac{Du}{Dt} \, d\mathbf{x} \cdot dt. \tag{5.12}$$

The material derivative $\frac{Du}{Dt}$ is written, when using a fixed reference system

$$\frac{Du}{Dt} = \frac{\partial u}{\partial t} + \mathbf{v} \cdot \nabla u. \tag{5.13}$$

However, when the reference system follows matter, the advective term can be discretized along the characteristic lines according to:

$$\frac{Du}{Dt} \approx \frac{u(\mathbf{x}, t) - \hat{u}(\mathbf{x}, t)}{\Delta t}, \tag{5.14}$$

where

$$\hat{u}(\mathbf{x}, t) = u(\mathbf{x} - \mathbf{v} \cdot \Delta t, t - \Delta t), \tag{5.15}$$

and $\mathbf{x} - \mathbf{v} \cdot \Delta t$ represents the root of the characteristic line at the previous time step, and hence the advantages of using an updated Lagrangian frame of reference.

Thus, if the time interval $\mathscr{I}$ is decomposed in P time steps of length Δt, i.e., $\mathscr{I} = [0, P \cdot \Delta t]$, Eq. (5.12) reduces to:

$$\mathscr{A} \approx \sum_{p=1}^{p=P} \int_{\Omega(t_p)} u^* \cdot (u(\mathbf{x}, t_p) - \hat{u}(\mathbf{x}, t_p)) \, d\mathbf{x}, \tag{5.16}$$

that is composed of two terms:

$$\mathscr{A}_1 \approx \sum_{p=1}^{p=P} \int_{\Omega(t_p)} u^* \cdot u(\mathbf{x}, t_p) \, d\mathbf{x}, \tag{5.17}$$

and

$$\mathscr{A}_2 \approx \sum_{p=1}^{p=P} \int_{\Omega(t_p)} u^* \cdot \hat{u}(\mathbf{x}, t_p) \, d\mathbf{x}. \tag{5.18}$$

Considering the approximation given by Eq. (5.3) we get

$$\mathscr{A}_1 \approx \sum_{p=1}^{p=P} \mathbf{U}^{*^T} \cdot \left(\int_{\Omega(t_p)} \mathbf{N}^{t_p} \cdot \mathbf{N}^{t_p^T} \, d\mathbf{x} \right) \cdot \mathbf{U}^{t_p} = \sum_{p=1}^{p=P} \mathbf{U}^{*^T} \cdot \mathbf{M}(t_p) \cdot \mathbf{U}^{t_p}. \tag{5.19}$$

Since $\Omega(t)$ is known $\forall t$, we can easily evaluate the integral

$$\mathbf{M}_k = \int_{\Omega(t_k)} \mathbf{N}^{t_k} \cdot \mathbf{N}^{t_k^T} \, d\mathbf{x}, \tag{5.20}$$

at different times t_k, $k = 1, \ldots, Q$.

Again, from these integrals we can define a matrix $\mathbf{M}$

$$\mathbf{M} = \begin{pmatrix} \mathbf{M}_1 \; \mathbf{M}_2 \; \cdots \; \mathbf{M}_Q \end{pmatrix}, \tag{5.21}$$

that after applying a singular value decomposition (SVD) allows us to write

$$\mathbf{M}(t) = \int_{\Omega(t)} \mathbf{N}^t \cdot \mathbf{N}^{t^T} \, d\mathbf{x} \approx \sum_{j=1}^{j=m_2} F_j^a(t) \cdot \mathbf{E}_j^a, \tag{5.22}$$

with $m_2 < Q$ and $m_2 < N_n$.

Thus, the term $\mathscr{A}_1$ reads:

$$\mathscr{A}_1 \approx \sum_{p=1}^{p=P} \mathbf{U}^{*^T} \cdot \left(\sum_{j=1}^{j=m_2} F_j^a(t_p) \cdot \mathbf{E}_j^a \right) \cdot \mathbf{U}^{t_p}. \tag{5.23}$$

We come now back to the second contribution $\mathscr{A}_2$. Firstly we define the approximation of $\hat{u}(\mathbf{x}, t)$:

$$\hat{u}(\mathbf{x} \in \Omega(t), t) \approx \sum_{i=1}^{i=N_n} \hat{N}_i(\mathbf{x}) \cdot u_i^{t-\Delta t} = \hat{\mathbf{N}}^T \cdot \mathbf{U}^{t-\Delta t}, \tag{5.24}$$

from which the integral $\mathscr{A}_2$ now reads

$$\mathscr{A}_2 \approx \sum_{p=1}^{p=P} \mathbf{U}^{*^T} \cdot \left(\int_{\Omega(t_p)} \mathbf{N}^{t_p} \cdot \hat{\mathbf{N}}^T \, d\mathbf{x} \right) \cdot \mathbf{U}^{t_p-\Delta t} = \sum_{p=1}^{p=P} \mathbf{U}^{*^T} \cdot \hat{\mathbf{M}}(t_p) \cdot \mathbf{U}^{t_p-\Delta t}. \tag{5.25}$$

Because $\Omega(t)$ is known $\forall t$, we can evaluate the integral

$$\hat{\mathbf{M}}_k = \int_{\Omega(t_k)} \mathbf{N}^{t_k} \cdot \hat{\mathbf{N}}^T \, d\mathbf{x}, \tag{5.26}$$

at different times t_k, $k = 1, \ldots, Q$.

From these integrals we could define a matrix $\hat{\mathbf{M}}$

$$\hat{\mathbf{M}} = \left(\hat{\mathbf{M}}_1, \hat{\mathbf{M}}_2, \ldots, \hat{\mathbf{M}}_Q \right), \tag{5.27}$$

that by applying a singular value decomposition (SVD) allows us to write

$$\hat{\mathbf{M}}(t) = \int_{\Omega(t)} \mathbf{N}^t \cdot \hat{\mathbf{N}}^T \, d\mathbf{x} \approx \sum_{j=1}^{j=m_3} F_j^u(t) \cdot \mathbf{E}_j^u, \tag{5.28}$$

with $m_3 < Q$ and $m_3 < N_n$.

Thus, the term $\mathscr{A}_2$ reads:

$$\mathscr{A}_2 \approx \sum_{p=1}^{p=P} \mathbf{U}^{*^T} \cdot \left(\sum_{j=1}^{j=m_3} F_j^u(t_p) \cdot \mathbf{E}_j^u \right) \cdot \mathbf{U}^{t_p-\Delta t}. \tag{5.29}$$

5.2.3 Source Term

We consider the source term in Eq. (5.2):

$$\mathscr{S} = \int_{\mathscr{I}} \int_{\Omega(t)} u^* \cdot s(\mathbf{x}, t) \, d\mathbf{x} \cdot dt. \tag{5.30}$$

By approximating the source term from:

$$s(\mathbf{x} \in \Omega(t), t) \approx \sum_{i=1}^{i=N_n} N_i^t(\mathbf{x}) \cdot s_i^t = \mathbf{N}^{t^T} \cdot \mathbf{S}^t, \tag{5.31}$$

the source term in the weak form of the problem can be written as:

$$\mathscr{S} = \int_{\mathscr{I}} \int_{\Omega(t)} \mathbf{U}^{*^T} \cdot \mathbf{N}^t \cdot \mathbf{N}^{t^T} \cdot \mathbf{S}^t \, d\mathbf{x} \cdot dt \tag{5.32}$$

$$= \int_{\mathscr{I}} \mathbf{U}^{*^T} \cdot \left(\int_{\Omega(t)} \mathbf{N}^t \cdot \mathbf{N}^{t^T} \, d\mathbf{x} \right) \cdot \mathbf{S}^t \, dt = \int_{\mathscr{I}} \mathbf{U}^{*^T} \cdot \mathbf{M}(t) \cdot \mathbf{S}^t \, dt, \tag{5.33}$$

which, considering the previous developments, results in:

$$\mathscr{S} = \int_{\mathscr{I}} \int_{\Omega(t)} \mathbf{U}^{*^T} \cdot \mathbf{N}^t \cdot \mathbf{N}^{t^T} \cdot \mathbf{S}^t \, d\mathbf{x} \cdot dt = \int_{\mathscr{I}} \mathbf{U}^{*^T} \cdot \left(\sum_{j=1}^{j=m_2} F_j^a(t) \cdot \mathbf{E}_j^a \right) \cdot \mathbf{S}^t \, dt. \tag{5.34}$$

By applying again a singular value decomposition $\mathbf{S}^t$ can be expressed in a separated form:

$$\mathbf{S}^t \approx \sum_{l=1}^{l=L} C_l(t) \cdot \mathbf{D}_l, \tag{5.35}$$

leading to:

$$\mathscr{S} = \int_{\mathscr{I}} \mathbf{U}^{*^T} \cdot \left(\sum_{j=1}^{j=m_2} F_j^a(t) \cdot \mathbf{E}_j^a \right) \cdot \left(\sum_{l=1}^{l=L} C_l(t) \cdot \mathbf{D}_l \right) dt. \tag{5.36}$$

5.3 Building-Up a Separated Representation of the Model Solution

Once the problem has been stated in a separated form, by applying SVD to every term in its weak form, the technique here proposed proceeds by constructing a separated, space-time, representation for the solution, $u = u(\mathbf{x}, t)$. In the mentioned separated representation, the model reads:

$$\sum_{p=1}^{k=P} \mathbf{U}^{*^T} \cdot \left(\sum_{j=1}^{j=m_2} F_j^a(t_p) \cdot \mathbf{E}_j^a \right) \cdot \mathbf{U}^{t_p}$$

$$- \sum_{p=1}^{p=P} \mathbf{U}^{*^T} \cdot \left(\sum_{j=1}^{j=m_3} F_j^u(t_p) \cdot \mathbf{E}_j^u \right) \cdot \mathbf{U}^{t_p - \Delta t}$$

$$+ \int_{\mathscr{I}} k \cdot \mathbf{U}^{*^T} \cdot \left(\sum_{j=1}^{j=m_1} F_j^d(t) \cdot \mathbf{E}_j^d \right) \cdot \mathbf{U}^t \, dt$$

$$- \int_{\mathscr{I}} \mathbf{U}^{*^T} \cdot \left(\sum_{j=1}^{j=m_2} F_j^a(t) \cdot \mathbf{E}_j^a \right) \cdot \mathbf{S}^t \, dt = 0. \tag{5.37}$$

Assuming that the model solution accepts a separated space-time representation, it is possible to look for an a priori separated representation of $\mathbf{U}^t$:

$$\mathbf{U}^t \approx \sum_{i=1}^{i=N} T_i(t) \cdot \mathbf{X}_i. \tag{5.38}$$

To construct such an approximation we proceed by computing a term of the finite sum at each iteration. Thus, we assume at iteration n that the first n terms of the sum have been already calculated, from which we can write the n-th order approximation of $\mathbf{U}^t$

$$\mathbf{U}^t \approx \sum_{i=1}^{i=n} T_i(t) \cdot \mathbf{X}_i. \tag{5.39}$$

At the following iteration, $n + 1$, we are looking for the new functional product $T_{n+1}(t) \cdot \mathbf{X}_{n+1}$. For the sake of simplicity functions $T_{n+1}(t)$ and $\mathbf{X}_{n+1}$ will be denoted by Υ and $\mathbf{R}$ respectively, where the dependence on t of Υ is omitted for the sake of clarity.

Thus, the $(n + 1)$-th order approximation reads:

$$\mathbf{U}^t \approx \sum_{i=1}^{i=n} T_i(t) \cdot \mathbf{X}_i + \Upsilon \cdot \mathbf{R}. \tag{5.40}$$

To compute both functions Υ and $\mathbf{R}$ we consider Eq. (5.37), where the trial function is given by (5.40) and the test function by:

$$\mathbf{U}^* = \Upsilon^* \cdot \mathbf{R} + \Upsilon \cdot \mathbf{R}^*. \tag{5.41}$$

Since the resulting problem is nonlinear, because of the product of both unknown functions Υ and $\mathbf{R}$, a linearization is therefore compulsory. The simplest one consists, as explained before, of a fixed point, alternating directions, strategy that computes $\mathbf{R}$ by assuming known Υ, then Υ from the just updated $\mathbf{R}$. Both steps are repeated until reaching the fixed point of both Υ and $\mathbf{R}$.

In what follows we detail both steps.

5.3.1 Computing the Space Function R

When Υ is assumed known, the test function $\mathbf{U}^*$ reduces to $\mathbf{U}^* = \Upsilon \cdot \mathbf{R}^*$. In this case the integral form reads:

$$\begin{aligned}
&\sum_{p=1}^{p=P} \mathbf{R}^{*^T} \cdot \Upsilon(t_p) \cdot \left(\sum_{j=1}^{j=m_2} F_j^a(t_p) \cdot \mathbf{E}_j^a \right) \cdot \left(\sum_{i=1}^{i=n} T_i(t_p) \cdot \mathbf{X}_i + \Upsilon(t_p) \cdot \mathbf{R} \right) \\
&\quad - \sum_{p=1}^{p=P} \mathbf{R}^{*^T} \cdot \Upsilon(t_p) \cdot \left(\sum_{j=1}^{j=m_3} F_j^u(t_p) \cdot \mathbf{E}_j^u \right) \cdot \left(\sum_{i=1}^{i=n} T_i(t_{p-1}) \cdot \mathbf{X}_i + \Upsilon(t_{p-1}) \cdot \mathbf{R} \right) \\
&\quad + \int_{\mathcal{I}} \mathbf{R}^{*^T} \cdot k \cdot \Upsilon(t) \cdot \left(\sum_{j=1}^{j=m_1} F_j^d(t) \cdot \mathbf{E}_j^d \right) \cdot \left(\sum_{i=1}^{i=n} T_i(t) \cdot \mathbf{X}_i + \Upsilon(t) \cdot \mathbf{R} \right) dt \\
&\quad - \int_{\mathcal{I}} \mathbf{R}^{*^T} \cdot \Upsilon(t) \cdot \left(\sum_{j=1}^{j=m_2} F_j^a(t) \cdot \mathbf{E}_j^a \right) \cdot \left(\sum_{l=1}^{l=L} C_l(t) \cdot \mathbf{D}_l \right) dt = 0 \qquad (5.42)
\end{aligned}$$

where $t_{p-1} = t_p - \Delta t$.

By using a simple numerical quadrature, the previous equation becomes:

$$\begin{aligned}
&\sum_{p=1}^{p=P} \mathbf{R}^{*^T} \cdot \Upsilon(t_p) \cdot \left(\sum_{j=1}^{j=m_2} F_j^a(t_p) \cdot \mathbf{E}_j^a \right) \cdot \left(\sum_{i=1}^{i=n} T_i(t_p) \cdot \mathbf{X}_i + \Upsilon(t_p) \cdot \mathbf{R} \right) \\
&\quad - \sum_{p=1}^{p=P} \mathbf{R}^{*^T} \cdot \Upsilon(t_p) \cdot \left(\sum_{j=1}^{j=m_3} F_j^u(t_p) \cdot \mathbf{E}_j^u \right) \cdot \left(\sum_{i=1}^{i=n} T_i(t_{p-1}) \cdot \mathbf{X}_i + \Upsilon(t_{p-1}) \cdot \mathbf{R} \right) \\
&\quad + \sum_{p=1}^{p=P} \mathbf{R}^{*^T} \cdot k \cdot \Upsilon(t_p) \cdot \left(\sum_{j=1}^{j=m_1} F_j^d(t_p) \cdot \mathbf{E}_j^d \right) \cdot \left(\sum_{i=1}^{i=n} T_i(t_p) \cdot \mathbf{X}_i + \Upsilon(t_p) \cdot \mathbf{R} \right) \cdot \Delta t \\
&\quad - \sum_{p=1}^{p=P} \mathbf{R}^{*^T} \cdot \Upsilon(t_p) \cdot \left(\sum_{j=1}^{j=m_2} F_j^a(t_p) \cdot \mathbf{E}_j^a \right) \cdot \left(\sum_{l=1}^{l=L} C_l(t_p) \cdot \mathbf{D}_l \right) \cdot \Delta t = 0. \qquad (5.43)
\end{aligned}$$

Because of the arbitrariness of $\mathbf{R}^*$, after developing all the calculations, Eq. (5.43) results in a linear system:

$$\mathbf{H} \cdot \mathbf{R} = \mathbf{Z}, \qquad (5.44)$$

from which we can update vector $\mathbf{R}$.

5.3.2 Computing the Time Function $\Upsilon(t)$

When $\mathbf{R}$ is assumed known, the test function $\mathbf{U}^*$ reduces to $\mathbf{U}^* = \Upsilon^* \cdot \mathbf{R}$. In this case is easy to verify that the discrete form reads:

$$\begin{aligned}
&\sum_{p=1}^{p=P} \Upsilon^*(t_p) \cdot \mathbf{R}^T \cdot \left(\sum_{j=1}^{j=m_2} F_j^a(t_p) \cdot \mathbf{E}_j^a \right) \cdot \left(\sum_{i=1}^{i=n} T_i(t_p) \cdot \mathbf{X}_i + \Upsilon(t_p) \cdot \mathbf{R} \right) \\
&\quad - \sum_{p=1}^{p=P} \Upsilon^*(t_p) \cdot \mathbf{R}^T \cdot \left(\sum_{j=1}^{j=m_3} F_j^u(t_p) \cdot \mathbf{E}_j^u \right) \cdot \left(\sum_{i=1}^{i=n} T_i(t_{p-1}) \cdot \mathbf{X}_i + \Upsilon(t_{p-1}) \cdot \mathbf{R} \right) \\
&\quad + \sum_{p=1}^{p=P} \Upsilon^*(t_p) \cdot \mathbf{R}^T \cdot k \cdot \left(\sum_{j=1}^{j=m_1} F_j^d(t_p) \cdot \mathbf{E}_j^d \right) \cdot \left(\sum_{i=1}^{i=n} T_i(t_p) \cdot \mathbf{X}_i + \Upsilon(t_p) \cdot \mathbf{R} \right) \cdot \Delta t \\
&\quad - \sum_{p=1}^{p=P} \Upsilon^*(t_p) \cdot \mathbf{R}^T \cdot \left(\sum_{j=1}^{j=m_2} F_j^a(t_p) \cdot \mathbf{E}_j^a \right) \cdot \left(\sum_{l=1}^{l=L} C_l(t_p) \cdot \mathbf{D}_l \right) \cdot \Delta t = 0.
\end{aligned} \tag{5.45}$$

Since we are assuming homogeneous initial condition, this results in $\Upsilon(t_0) = 0$. Then, the arbitrariness of $\Upsilon^*(t_p)$, $\forall p \geq 1$ implies:

$$\begin{aligned}
&\mathbf{R}^T \cdot \left(\sum_{j=1}^{j=m_2} F_j^a(t_p) \cdot \mathbf{E}_j^a \right) \cdot \left(\sum_{i=1}^{i=n} T_i(t_p) \cdot \mathbf{X}_i + \Upsilon(t_p) \cdot \mathbf{R} \right) \\
&\quad - \mathbf{R}^T \cdot \left(\sum_{j=1}^{j=m_3} F_j^u(t_p) \cdot \mathbf{E}_j^u \right) \cdot \left(\sum_{i=1}^{i=n} T_i(t_{p-1}) \cdot \mathbf{X}_i + \Upsilon(t_{p-1}) \cdot \mathbf{R} \right) \\
&\quad + \mathbf{R}^T \cdot k \cdot \left(\sum_{j=1}^{j=m_1} F_j^d(t_p) \cdot \mathbf{E}_j^d \right) \cdot \left(\sum_{i=1}^{i=n} T_i(t_p) \cdot \mathbf{X}_i + \Upsilon(t_p) \cdot \mathbf{R} \right) \cdot \Delta t \\
&\quad - \mathbf{R}^T \cdot \left(\sum_{j=1}^{j=m_2} F_j^a(t_p) \cdot \mathbf{E}_j^a \right) \cdot \left(\sum_{l=1}^{l=L} C_l(t_p) \cdot \mathbf{D}_l \right) \cdot \Delta t = 0,
\end{aligned} \tag{5.46}$$

that after making the indicated calculation results in a simple linear equation for each t_p:

$$\Upsilon(t_p) = a_p \cdot \Upsilon(t_{p-1}) + b_p, \quad \forall p \geq 1. \tag{5.47}$$

5.4 Numerical Test

In this section we consider a numerical example consisting of a solid workpiece occupying at $t = 0$ the domain $\Omega^0 = (-0.5, 0.5) \times (0, 5)$. The piece is being compressed from its upper face. The domain evolves consequently to take intermediate configurations $\Omega(t)$ until reaching its final geometry at time $t = 1.28$. The tool of unit length $(-0.5, 0.5)$ is compressing the workpiece at a constant compression velocity. Assuming known the geometry evolution $\Omega(t)$ we solve the thermal model defined in $\Omega(t)$ in a non-incremental way. We consider the following initial and boundary conditions:

$$\begin{cases} u(\mathbf{x} \in \Omega^0, t = 0) = 1 \\ u(\mathbf{x} \in \Gamma_c(t), t) = 0 \\ \nabla u(\mathbf{x} \in \Gamma_f(t), t) \cdot \mathbf{n} = 0 \end{cases}, \tag{5.48}$$

where $\Gamma_c(t)$ and $\Gamma_f(t)$ represent the parts of the boundary of $\Omega(t)$, $\Gamma(t) \equiv \partial\Omega(t)$, in contact with the tool or the work plane $y = 0$ and the free boundary respectively.

After applying the strategy described in the previous section, $N = 15$ modes were found to be enough for representing the whole thermal history $u(\mathbf{x} \in \Omega(t), t)$:

$$\mathbf{U}^t \approx \sum_{i=1}^{i=15} T_i(t) \cdot \mathbf{X}_i. \tag{5.49}$$

Functions T_i were computed only at times t_p and $\mathbf{X}_i$ consists of a vector containing the nodal values related to any nodal distribution $\tilde{\mathbf{x}}_j^t$ in $\Omega(t)$.

For the sake of clarity we defined functions $G_i(t)$ by interpolating $T_i(t_p)$ values and defined functions $F_i(\mathbf{x} \in \Omega^0)$ by interpolating values in $\mathbf{X}_i$ on the initial configuration Ω^0. Figure 5.1 depicts the five most significant space modes $F_i, i = 1, \ldots, 5$; as well as $G_i(t), i = 1, \ldots, 15$.

Now, we are in the position of assigning vectors $\mathbf{X}_i$ to nodes $\tilde{\mathbf{x}}_j^t$ related to the configuration $\Omega(t)$ and then to reconstruct the solution in $\Omega(t)$. Figure 5.2 depicts the reconstructed temperature field in $\Omega(t)$ for six different time instants.

Figure 5.3 depicts, in turn, the error of the approximation for different numbers of terms in the approximation (ranging from 10 to 30) versus the incremental, standard, finite element solution of the problem.

5.5 Towards Parametric Modeling in Evolving Domains

The extension of the previously introduced technique to the case of parametric models on evolving domains is straightforward. To this end we should come back to Sect. 5.3 and consider the parametric dependency of u on k, looking for the separated representation:

$$\mathbf{U}_k^t \approx \sum_{i=1}^{i=N} T_i(t) \cdot K_i(k) \cdot \mathbf{X}_i. \tag{5.50}$$

For constructing such an approximation we proceed by computing a term of the finite sum at each iteration. Thus, we assume at iteration n that the first n terms of the sum have been already calculated, from which we can write the n-order approximation

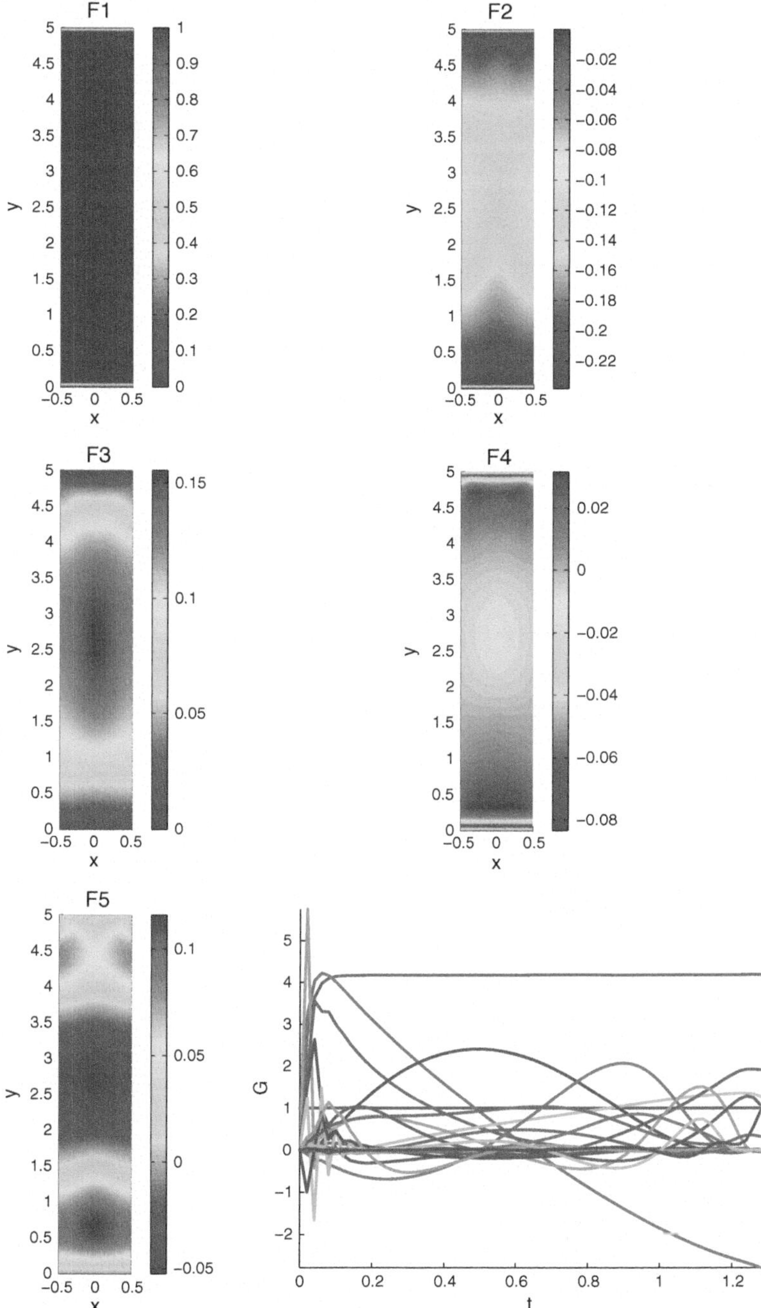

Fig. 5.1 Space and time functions involved in the separated representation of $u(\mathbf{x} \in \Omega(t), t)$

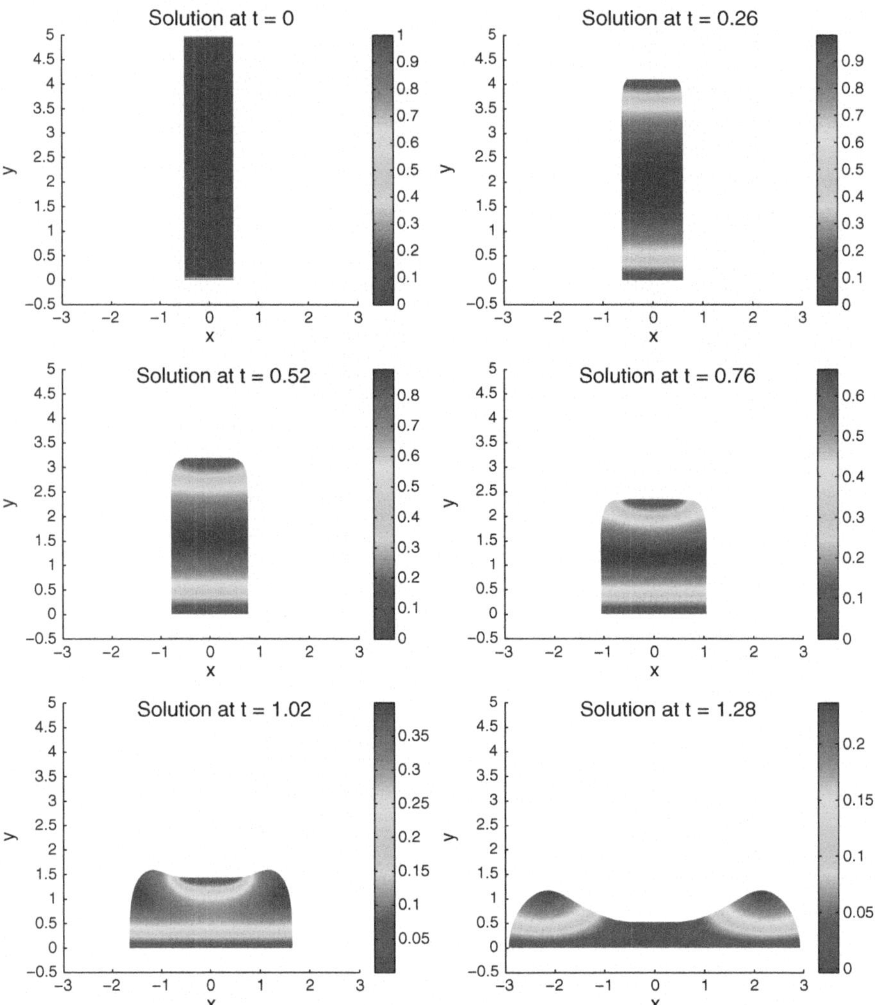

Fig. 5.2 Reconstructed temperature field $u(\mathbf{x} \in \Omega(t), t)$ at six different instants

$$\mathbf{U}_k^t \approx \sum_{i=1}^{i=n} T_i(t) \cdot K_i(k) \cdot \mathbf{X}_i. \tag{5.51}$$

Now, at the next iteration $n+1$ we look for the new functional product $T_{n+1}(t) \cdot K_{n+1} \cdot \mathbf{X}_{n+1}$. For the sake of simplicity functions $T_{n+1}(t)$, K_{n+1} and $\mathbf{X}_{n+1}$ will be noted by Υ, W and $\mathbf{R}$ respectively, where the dependence on t of Υ, and on k of W is omitted for the sake of clarity.

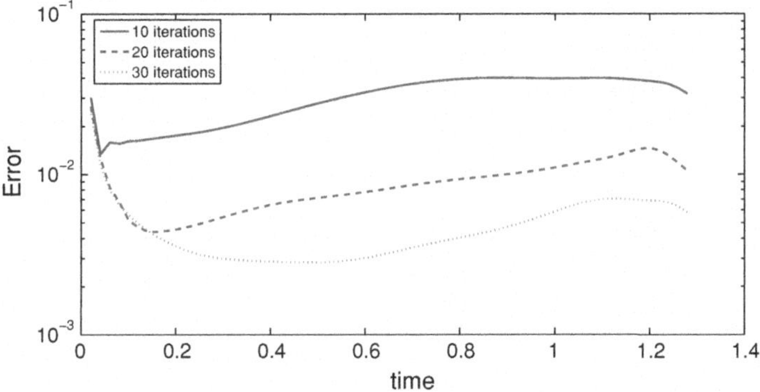

Fig. 5.3 Error in the approximation for different number of terms in the sum. A standard, incremental finite element solution has been taken as reference

Thus, the $(n+1)$-th order approximation reads:

$$\mathbf{U}_k^t \approx \sum_{i=1}^{i=n} T_i(t) \cdot K_i(k) \cdot \mathbf{X}_i + \Upsilon \cdot W \cdot \mathbf{R}. \tag{5.52}$$

For computing functions Υ, W and $\mathbf{R}$ we consider Eq. (5.37) where the trial function is given by (5.52) and the test function by:

$$\mathbf{U}^* = \Upsilon^* \cdot W \cdot \mathbf{R} + \Upsilon \cdot W^* \cdot \mathbf{R} + \Upsilon \cdot W \cdot \mathbf{R}^*. \tag{5.53}$$

Since the resulting problem is nonlinear because of the product of the three unknown functions Υ, W and $\mathbf{R}$ a linearization is compulsory. The simplest one consists of the fixed point alternating directions strategy presented above. By generalizing the procedure previously described we can compute the parametric and non-incremental separated representation. In the next section we consider the parametric solution of the problem just considered in the previous section.

5.6 Numerical Test Involving Parametric Modeling

In this section we consider the problem analyzed in Sect. 4 where the material conductivity k is now considered as a model extra-coordinate taking values in the interval $k \in \Im = (0, 1)$. The strategy is now a mere combination of those applied for parametric problems in steady domains, and that for standard problems in evolving domains.

Figure 5.4 depicts the four most significant space modes F_i, $i = 1, \ldots, 4$, where again F_i refers to the interpolation defined from values in $\mathbf{X}_i$ on the initial configuration Ω^0.

Figure 5.5 depicts the most significant functions depending on time and on conductivity.

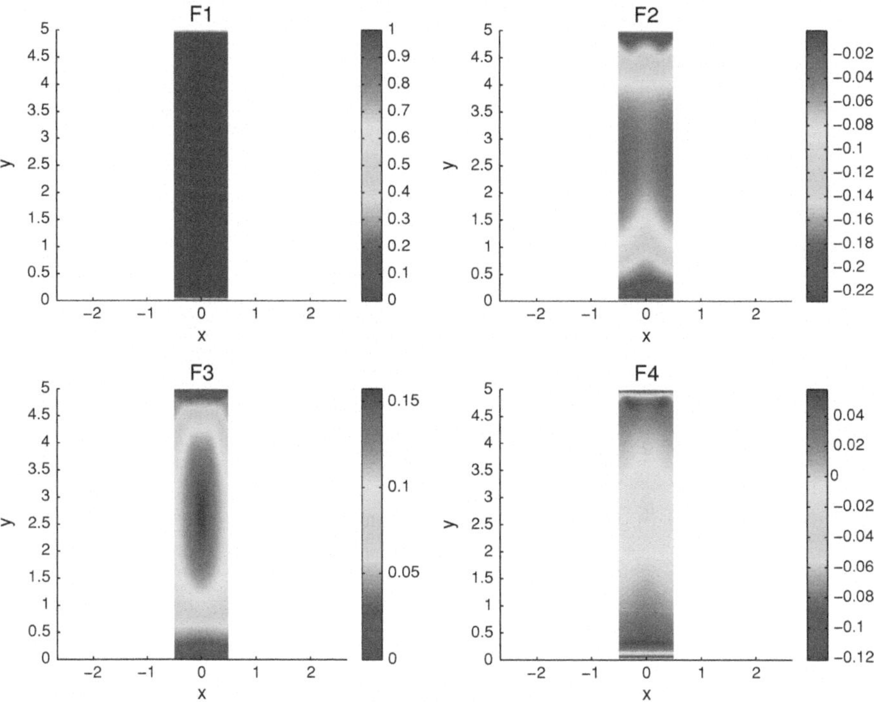

Fig. 5.4 Four most significant space functions involved in the separated representation of $u(\mathbf{x} \in \Omega(t), t, k)$

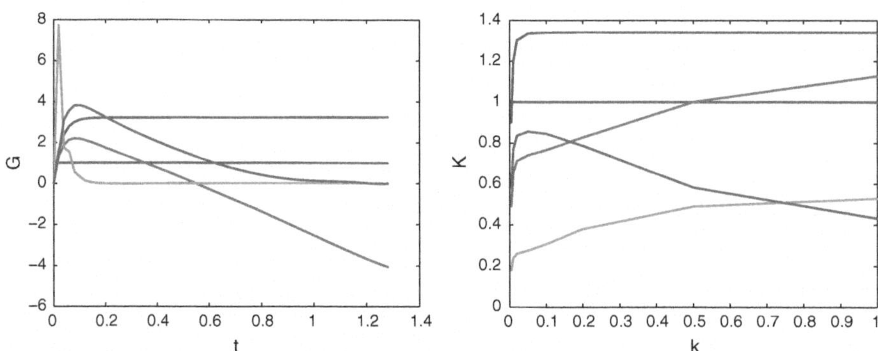

Fig. 5.5 Most significant functions depending on the space and conductivity involved in the separated representation of $u(\mathbf{x} \in \Omega(t), t, k)$

Finally, Fig. 5.6 depicts the temperature field reconstructed at the final geometry $\Omega(t = 1.28)$ for different values of the thermal conductivity. It can be noticed that the higher the conductivity the faster is the cooling process induced by the lower and

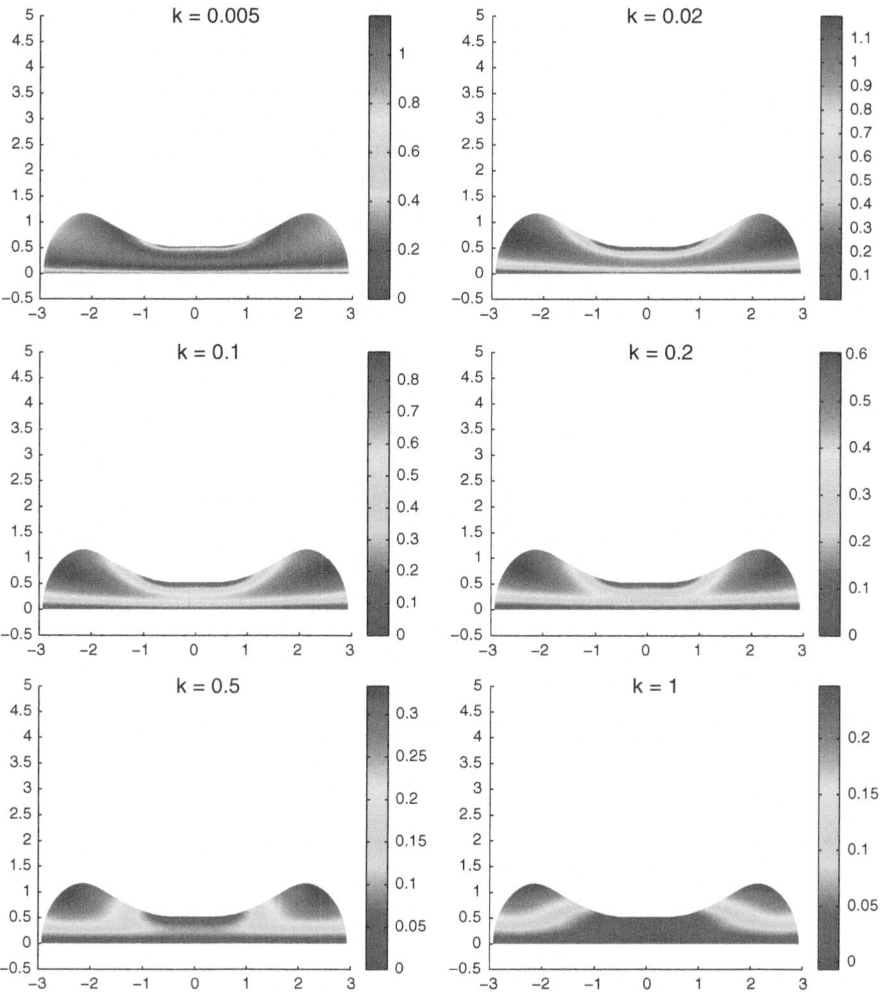

Fig. 5.6 Reconstructed temperature field $u(\mathbf{x} \in \Omega(t = 1.28), t, k)$ for different values of the thermal conductivity k

constant temperatures enforced on the tool and working plane surfaces in contact with the workpiece.

5.7 Conclusions

A novel strategy for a priori construction of reduced bases for problems defined in evolving domains is presented. The main challenge in this class of problems derives precisely from the deformation of the problem domain, which prevents the direct application of classical, a posteriori, techniques such as proper orthogonal decomposition, to obtain an appropriate reduced basis. The evolving nature of the domain obscures the concept of a snapshot of the system state, requiring specific treatments.

However, it has been demonstrated that a combination of an updated Lagrangian approach for the description of domain's kinematics and a PGD-based obtention of the set of reduced basis in a separated space-time (possibly space-parameters-time) representation gives a very convenient way of constructing reduced basis. This basis can be advantageously employed to simulate complex problems at a very reduced CPU cost, as proven in the vast corps of literature devoted to this end.

References

1. J. Donea, A. Huerta, *Finite Element Methods for Flow Problems* (John Wiley and Sons, Chichester, 2002)
2. R.W. Lewis, S.E. Navti, C. Taylor, A mixed lagrangian-eulerian approach to modelling fuid fow during mould filling. Int. J. Numer. Meth. Fluids **25**, 931–952 (1997)
3. D. Ryckelynck, F. Chinesta, E. Cueto, A. Ammar, On the a priori model reduction: overview and recent developments. Arch. Comput. Methods Eng. State Art Rev. **13**(1), 91–128 (2006)
4. F. Chinesta, S. Cescotto, E. Cueto, Ph. Lorong, *Natural Element Method for the Simulation of Structures and Processes* (Iste-Wiley, London, 2011)
5. P. Ladeveze, *Nonlinear Computational Structural Mechanics* (Springer, NY, 1999)
6. D. Gonzalez, A. Ammar, F. Chinesta, E. Cueto, Recent advances in the use of separated representations. Int. J. Numer. Meth. Eng. **81**(5), 637–659 (2010)

Chapter 6
Space Separation

The solution of 3D models in degenerated geometries in which some characteristic dimensions are much lower than the other ones—e.g. beams, plates or shells—is a tricky issue when using standard mesh-based discretization techniques. Separated representations allow decoupling the meshes used for approximating the solution along each coordinate. Thus, in plate or shell geometries 3D solutions can be obtained from a sequence of 2D and 1D problems allowing a fine and accurate representation of the solution evolution along the thickness coordinate while keeping the computational complexity characteristic of 2D simulations. In this chapter we revisit the application of such methodology for addressing different physics (thermal models, solid and fluid mechanics and electromagnetic problems) in such degenerated geometries.

6.1 In-Plane/Out-of-Plane Separated Representation

Plates and shells are very common in nature and thus they inspired engineers that used both from the very beginning of structural mechanics. In general the design of such structural elements requires the calculation of stresses, strains and displacements for the design loads. Strains and stresses are related by the so-called constitutive law, the simplest one being the linear elasticity. Moreover other physics can be encountered in these structural elements as for example thermal processes, electromagnetism or fluid flows. Typically composites manufacturing processes usually involve a resin flowing into a double scale porous media, that polymerizes by a thermal, chemical or electromagnetic activation. Welding processes could also involve the use of thermal, mechanical or electromagnetic devices.

Thus, structural or processes design always involve the solution of a set of partial differential equations in the degenerate domain of the plate or the shell with appropriate initial and boundary conditions. These domains are degenerated because one of their characteristic dimensions (thickness in the present case) is much lower than the other characteristic dimensions. We will understand the consequences of such

F. Chinesta and E. Cueto, *PGD-Based Modeling of Materials, Structures and Processes*, ESAFORM Bookseries on Material Forming, DOI: 10.1007/978-3-319-06182-5_6,

degeneracy later. When analytical solutions are neither available nor possible because of geometrical or behavior complexities, the solution must be calculated by invoking any of the available numerical techniques (finite elements, finite differences, finite volumes, methods of particles, ...).

In the numerical framework the solution will be only obtained in a discrete number of points, usually called nodes, properly distributed in the domain. From the solution at those points, it can be interpolated at any other point in the domain. In general regular nodal distributions are preferred because they offer the best accuracy. In the case of degenerated plate or shell domains one could expect that if the solution evolves significantly in the thickness direction, a large enough number of nodes must be distributed along the thickness direction to ensure the accurate representation of the field evolution in that direction. In that case, a regular nodal distribution in the whole domain will imply the use of an extremely large number of nodes with a consequent impact on the numerical solution efficiently.

When simple behaviors and domains were considered, plate and shell theories were developed allowing, through the introduction of some hypotheses, reducing the 3D complexity to the 2D related to the problem now formulated by considering the in-plane coordinates. The use of these theories was extended gradually for addressing larger and more complex geometries (anisotropic laminates, ...) and behaviors. These simplified "plate models" exist for most of the physics just cited, but their applicability is in many cases too constrained.

Moreover, as soon as richer physics are involved in the models and the considered geometries differ from those ensuring the validity of the different reduction hypotheses, efficient simulations are compromised. For example in large composite part manufacturing processes, many reactions and thermal processes inducing significant evolutions on the thermomechanical fields in the thickness occur. These inhomogeneities are at the origin of residual stresses and the associated distortion of the formed parts.

In these circumstances as just indicated the reduction from the 3D model to a 2D simplified one is not obvious, and 3D simulations appear many times as the only valid route for addressing such models, that despite the fact of being defined in degenerated geometries (plate or shell) they seem to require a fully 3D solution. In order to integrate such calculations (fully 3D and implying an impressive number of degrees of freedom) in usual design procedures, a new efficient (fast and accurate) solution procedure is needed.

A direct consequence of the separated representations involved in the PGD is space separation. Thus an in-plane-out-of-plane decomposition was proposed for solving flow problems in laminates [1, 2], thermal problems in extruded geometries [3] and laminates [4], elasticity problems in plates [5] and shells geometries [6, 7]. In general the 3D solution was obtained from the solution of a sequence of 2D problems (the ones involving the in-plane coordinates) and 1D problems (the ones involving the coordinate related to the plate thickness).

It is important to emphasize the fact that these approaches are radically different from standard plate and shell approaches. We proposed a 3D solver able to compute the different unknown fields without the necessity of introducing any hypothesis.

The most outstanding advantage is that 3D solutions can be obtained with a computational cost characteristic of standard 2D solutions.

In this work we revisit the in-plane-out-of-plane representation and its application for addressing different physics (thermal, elastic, flow and electromagnetic models) in plate domains, eventually consisting of a laminate composed of several anisotropic plies.

6.1.1 Elastic Problem in Plate Domains

We proposed in [5] an original in-plane/out-of-plane decomposition of the 3D elastic solution in a plate geometry. The elastic problem was defined in a plate domain $\Xi = \Omega \times \mathscr{I}$ with $\mathbf{x} = (x_1, x_2) \in \Omega$, $\Omega \subset \mathscr{R}^2$ and $x_3 \in \mathscr{I}$, $\mathscr{I} = [0, H] \subset \mathscr{R}$, being H the plate thickness. The separated representation of the displacement field $\mathbf{u} = (u_1, u_2, u_3)$ reads:

$$\mathbf{u}(x_1, x_2, x_3) = \begin{pmatrix} u_1(x_1, x_2, x_3) \\ u_2(x_1, x_2, x_3) \\ u_3(x_1, x_2, x_3) \end{pmatrix} \approx \sum_{i=1}^{N} \begin{pmatrix} P_1^i(x_1, x_2) \cdot T_1^i(x_3) \\ P_2^i(x_1, x_2) \cdot T_2^i(x_3) \\ P_3^i(x_1, x_2) \cdot T_3^i(x_3) \end{pmatrix}, \tag{6.1}$$

where $P_k^i, k = 1, 2, 3$, are functions of the in-plane coordinates $\mathbf{x} = (x_1, x_2)$ whereas T_k^i, $k = 1, 2, 3$, are functions involving the thickness coordinate x_3.

Expression (6.1) can be written in a more compact form by using the Hadamard product:

$$\mathbf{u}(\mathbf{x}, x_3) \approx \sum_{i=1}^{N} \mathbf{P}^i(\mathbf{x}) \circ \mathbf{T}^i(x_3), \tag{6.2}$$

where vectors $\mathbf{P}^i$ and $\mathbf{T}^i$ contain functions P_k^i and T_k^i respectively.

Because neither the number of terms in the separated representation of the displacement field nor the dependence on x_3 of functions T_k^i are assumed *a priori*, the approximation is flexible enough to represent the fully 3D solution, being obviously more general than theories assuming particular *a priori* evolutions in the thickness direction x_3.

Let us consider a linear elasticity problem on a plate domain $\Xi = \Omega \times \mathscr{I}$. The weak formulation reads:

$$\int_{\Xi} \varepsilon(\mathbf{u}^*)^T \cdot \mathbf{K} \cdot \varepsilon(\mathbf{u}) \, d\mathbf{x} = \int_{\Xi} \mathbf{u}^* \cdot \mathbf{f}_d \, d\mathbf{x} + \int_{\Gamma_N} \mathbf{u}^* \cdot \mathbf{F}_d \, d\mathbf{x}, \tag{6.3}$$

with $\mathbf{K}$ the generalized 6×6 Hooke tensor, $\mathbf{f}_d$ represents the volumetric body forces while $\mathbf{F}_d$ represents the traction applied on the boundary Γ_N. In what follows we assume that $\mathbf{K}$, $\mathbf{f}_d$ and $\mathbf{F}_d$ accepts an in-plane-out-of-plane separated representation (we come back to this issue later).

The separated representation constructor proceeds by computing a term of the sum at each iteration. Assuming that the first $n-1$ modes (terms of the finite sum) of the solution were already computed, $\mathbf{u}^{n-1}(\mathbf{x}, x_3)$ with $n \geq 1$, the solution enrichment reads:

$$\mathbf{u}^n(\mathbf{x}, x_3) = \mathbf{u}^{n-1}(\mathbf{x}, x_3) + \mathbf{P}^n(\mathbf{x}) \circ \mathbf{T}^n(x_3), \tag{6.4}$$

where both vectors $\mathbf{P}^n$ and $\mathbf{T}^n$ containing functions P_i^n and T_i^n ($i = 1, 2, 3$) depending on $\mathbf{x}$ and x_3 respectively, are unknown at the present iteration, resulting in a nonlinear problem.

We proceed by considering the simplest linearization strategy, an alternating directions fixed point algorithm, that proceeds by calculating $\mathbf{P}^n$ by assuming $\mathbf{T}^n$ known, and then by updating $\mathbf{T}^n$ from the just calculated $\mathbf{P}^n$. The iteration procedure continues until convergence, that is, until reaching the fixed point.

When $\mathbf{T}^n$ is assumed known, we consider the test function $\mathbf{u}^\star$ given by $\mathbf{P}^\star \circ \mathbf{T}^n$. By introducing the trial and test functions into the weak form and then integrating in $\mathcal{I}$ because all the functions depending on the thickness coordinate are known, we obtain a 2D weak formulation defined in Ω whose discretization (by using a standard discretization strategy, e.g. finite elements) allows computing $\mathbf{P}^n$.

Analogously, when $\mathbf{P}^n$ is assumed known, the test function $\mathbf{u}^\star$ is given by $\mathbf{P}^n \circ \mathbf{T}^\star$. By introducing the trial and test functions into the weak form and then integrating in Ω because all the functions depending on the in-plane coordinates $\mathbf{x}$ are at present known, we obtain a 1D weak formulation defined in $\mathcal{I}$ whose discretization (using any technique for solving standard ODE equations) allows computing $\mathbf{T}^n$.

As discussed in [5] this separated representation allows computing 3D solutions while keeping a computational complexity characteristic of 2D solution procedures. If we consider a hexahedral domain discretized using a regular structured grid with N_1, N_2 and N_3 nodes in the x_1, x_2 and x_3 directions respectively, usual mesh-based discretization strategies imply a challenging issue because the number of nodes involved in the model scales with $N_1 \cdot N_2 \cdot N_3$. However, by using the separated representation and assuming that the solution involves N modes, one must solve about N 2D problems related to the functions involving the in-plane coordinates $\mathbf{x}$ and the same number of 1D problems related to the functions involving the thickness coordinate x_3. The computing time related to the solution of the one-dimensional problems can be neglected with respect to the one required for solving the two-dimensional ones. Thus, the resulting complexity scales as $N \cdot N_1 \cdot N_2$. By comparing both complexities we can notice that as soon as $N_3 \gg N$ the use of separated representations leads to impressive computing time savings, making possible the solution of models never until now solved, and even using light computing platforms [5].

6.1.2 Elastic Problem in Shell Domains

The shell domain Ω^S, assumed with constant thickness, can be described from a reference surface $\mathbf{X}$, that in what follows will be identified to the shell middle surface but that in the general case could be any other one, parametrized by the coordinates ξ, η, that is $\mathbf{X}(\xi, \eta)$, where:

$$\mathbf{X}(\xi, \eta) = \begin{pmatrix} X_1(\xi, \eta) \\ X_2(\xi, \eta) \\ X_3(\xi, \eta) \end{pmatrix}. \tag{6.5}$$

Let $\mathbf{n}$ denote the unit vector normal to the middle surface, the shell domain Ω^S can be parametrized from:

$$\mathbf{x}(\xi, \eta, \zeta) = \mathbf{X}(\xi, \eta) + \zeta \cdot \mathbf{n}. \tag{6.6}$$

The geometrical transformation $(\xi, \eta, \zeta) \rightarrow (x_1, x_2, x_3)$ involves

$$\tilde{\mathbf{F}} = \left[\frac{\partial \mathbf{x}}{\partial \xi} \ \frac{\partial \mathbf{x}}{\partial \eta} \ \mathbf{n} \right]. \tag{6.7}$$

With the weak form of the elastic problem defined in the reference domain $\Xi = \Omega \times \mathscr{I}$, with $(\xi, \eta) \in \Omega$ and $\zeta \in \mathscr{I}$, the situation is quite similar to the one encountered in the analysis of elastic problems in plate geometries that was addressed in [5].

In Bognet et al. [7] we considered the in-plane-out-of-plane separated representation of the displacement field, similar to (6.1) but now involving the coordinates (ξ, η, ζ)

$$\mathbf{u}(\xi, \eta, \zeta) = \begin{pmatrix} u_1(\xi, \eta, \zeta) \\ u_2(\xi, \eta, \zeta) \\ u_3(\xi, \eta, \zeta) \end{pmatrix} \approx \sum_{i=1}^{N} \begin{pmatrix} P_1^i(\xi, \eta) \cdot T_1^i(\zeta) \\ P_2^i(\xi, \eta) \cdot T_2^i(\zeta) \\ P_3^i(\xi, \eta) \cdot T_3^i(\zeta) \end{pmatrix}, \tag{6.8}$$

or in a more compact form

$$\mathbf{u}(,, \eta, \zeta) \approx \sum_{i=1}^{N} \mathbf{P}^i(\xi, \eta) \circ \mathbf{T}^i(\zeta). \tag{6.9}$$

6.1.3 Darcy's Flow Model

We now illustrate the application of separated representations to the modeling of resin transfer molding processes comprehensively addressed in Chinesta, Ammar et al. [8]. We consider the flow within a porous medium in a plate domain $\Xi = \Omega \times \mathscr{I}$ with

$\Omega \subset \mathscr{R}^2$ and $\mathscr{I} = [0, H] \subset \mathscr{R}$. The governing equation is obtained by combining Darcy's law, which relates the fluid velocity to the pressure gradient

$$\mathbf{v} = -\mathbf{K} \cdot \nabla p, \tag{6.10}$$

and the incompressibility constraint,

$$\nabla \cdot \mathbf{v} = 0. \tag{6.11}$$

Introduction of Eq. (6.10) into Eq. (6.11) yields a single equation for the pressure field

$$\nabla \cdot (\mathbf{K} \cdot \nabla p) = 0. \tag{6.12}$$

Remark The heat equation being formally similar to Eq. (6.12) the considerations that follow also apply for the solution of thermal models in plate geometries. □

Again we assume the permeability separability

$$\mathbf{K}(x, y, z) = \sum_{i=1}^{P} \mathbf{K}_i(\mathbf{x}) \cdot \xi_i(z), \tag{6.13}$$

where $\mathbf{x}$ denotes the in-plane coordinates, i.e. $\mathbf{x} = (x, y) \in \Omega$.

The weak form of Eq. (6.12) reads:

$$\int_{\Xi} \nabla p^* \cdot (\mathbf{K} \cdot \nabla p) \; d\Xi = 0, \tag{6.14}$$

for all test functions p^* selected in an appropriate functional space. Dirichlet boundary conditions are imposed for the pressure at the inlet and outlet of the flow domain, while zero flux (i.e. no flow) is imposed elsewhere as a weak boundary condition. We seek an approximate solution $p(x, y, z)$ in the separated form:

$$p(\mathbf{x}, z) \approx \sum_{j=1}^{N} X_j(\mathbf{x}) \cdot Z_j(z). \tag{6.15}$$

The PGD algorithm then proceeds as follows. Assume that the first n functional products have been computed, i.e.,

$$p^n(\mathbf{x}, z) = \sum_{j=1}^{n} X_j(\mathbf{x}) \cdot Z_j(z), \tag{6.16}$$

is a known quantity. We must now perform an enrichment step to obtain

$$p^{n+1}(\mathbf{x}, z) = p^n(\mathbf{x}, z) + R(\mathbf{x}) \cdot S(z). \tag{6.17}$$

The test function involved in the weak form is given by

$$p^*(\mathbf{x}, z) = R^*(\mathbf{x}) \cdot S(z) + R(\mathbf{x}) \cdot S^*(z). \tag{6.18}$$

Introducing Eqs. (6.17) and (6.18) into Eq. (6.14), we obtain

$$\int_\Xi \left(\begin{pmatrix} \tilde{\nabla} R^* \cdot S \\ R^* \cdot \frac{dS}{dz} \end{pmatrix} + \begin{pmatrix} \tilde{\nabla} R \cdot S^* \\ R \cdot \frac{dS^*}{dz} \end{pmatrix} \right) \cdot \left(\mathbf{K} \cdot \begin{pmatrix} \tilde{\nabla} R \cdot S \\ R \cdot \frac{dS}{dz} \end{pmatrix} \right) d\Xi \tag{6.19}$$

$$= - \int_\Xi \left(\begin{pmatrix} \tilde{\nabla} R^* \cdot S \\ R^* \cdot \frac{dS}{dz} \end{pmatrix} + \begin{pmatrix} \tilde{\nabla} R \cdot S^* \\ R \cdot \frac{dS^*}{dz} \end{pmatrix} \right) \cdot \mathbf{Q}^n \, d\Xi,$$

where $\tilde{\nabla}$ denotes the plane component of the gradient operator, i.e., $\tilde{\nabla}^T = \left(\frac{\partial}{\partial x}, \frac{\partial}{\partial y} \right)$ and $\mathbf{Q}^n$ is a flux term known at step n:

$$\mathbf{Q}^n = \mathbf{K} \cdot \sum_{j=1}^{n} \begin{pmatrix} \tilde{\nabla} X_j(\mathbf{x}) \cdot Z_j(z) \\ X_j(\mathbf{x}) \cdot \frac{dZ_j(z)}{dz} \end{pmatrix}. \tag{6.20}$$

As discussed previously, each enrichment step of the PGD algorithm is a nonlinear problem which must be performed by means of a suitable iterative process. The simplest one proceeds assuming $R(\mathbf{x})$ known to obtain $S(z)$, and then updating $R(\mathbf{x})$. The process continues until reaching convergence. The converged solutions provide the next functional product of the PGD: $R(\mathbf{x}) \rightarrow X_{n+1}(\mathbf{x})$ and $S(z) \rightarrow Z_{n+1}(z)$.

6.1.4 Stokes Flow Model

The Stokes model is defined in $\Xi = \Omega \times \mathscr{I}$ and reads, for an incompressible fluid:

$$\begin{cases} \nabla p = \nabla \cdot (\eta \cdot \nabla \mathbf{v}) \\ \nabla \cdot \mathbf{v} = 0 \end{cases}. \tag{6.21}$$

To circumvent the issue related to stable mixed formulations (LBB conditions) within the separated representation we consider a penalty formulation that modifies the mass balance by introducing a penalty coefficient λ small enough

$$\nabla \cdot \mathbf{v} + \lambda \cdot p = 0, \tag{6.22}$$

or more explicitly

$$p = -\frac{1}{\lambda}\left(\frac{\partial u}{\partial x} + \frac{\partial v}{\partial y} + \frac{\partial w}{\partial z}\right) = -\frac{\nabla \cdot \mathbf{v}}{\lambda}. \tag{6.23}$$

By replacing it into the balance of momentum (first equation in (6.21)) we obtain

$$\nabla\left(\nabla \cdot \mathbf{v}\right) + \xi \Delta \mathbf{v} = 0, \tag{6.24}$$

with $\xi = \eta \cdot \lambda$.

In the present case it suffices to consider the separated representation of the velocity field according to (Ghnatios et al. [2]):

$$\mathbf{v} = \begin{pmatrix} u \\ v \\ w \end{pmatrix} \approx \begin{pmatrix} \sum_{i=1}^{i=N} X_i^u(x, y) \cdot Z_i^u(z) \\ \sum_{i=1}^{i=N} X_i^v(x, y) \cdot Z_i^v(z) \\ \sum_{i=1}^{i=N} X_i^w(x, y) \cdot Z_i^w(z) \end{pmatrix}, \tag{6.25}$$

that leads to a separated representation of the strain rate which, introduced into the Stokes problem weak form, allows the calculation of functions $X_i(x, y)$ by solving the corresponding 2D problems and functions $Z_i(z)$ by solving the associated 1D problems. Because of the one-dimensional character of problems defined in the laminate thickness we can use extremely fine descriptions along the thickness direction without a significant impact on the computational efficiency.

6.2 Laminates

In the case of laminates, several plies with different thermomechanical properties (eventually anisotropic) are found through the domain thickness. In the case of addressing the flow in a porous preform we could suppose P different anisotropic plies of thickness h, each one characterized by a permeability tensor $\mathbf{K}_i(x, y)$ that is assumed constant through the ply thickness. Then, we define a characteristic function

$$\xi_i(z) = \begin{cases} 1 & z_i \le z \le z_{i+1} \\ 0 & \text{otherwise} \end{cases}, \tag{6.26}$$

where $z_i = (i-1) \cdot h$ is the location of ply i in the plate thickness. The laminate's permeability is thus given in separated form

$$\mathbf{K}(x, y, z) = \sum_{i=1}^{P} \mathbf{K}_i(\mathbf{x}) \cdot \xi_i(z), \tag{6.27}$$

where $\mathbf{x}$ denotes the in-plane coordinates, i.e., $\mathbf{x} = (x, y) \in \Omega$.

6.2.1 Brinkman Model

In composites manufacturing processes, resin located among the fibers in the reinforcement layers also flows. A usual approach for evaluating the resin flow in such circumstances consists of solving the associated Darcy's model. It is well known that Darcy-Stokes coupling at the interlayers generates numerical instabilities because the localized boundary layers whose accurate description requires very rich representations (very fine meshes along the laminate thickness).

In Ghnatios et al. [2] we proposed the use of the Brinkman model that allows representing in an unified manner both the Darcy and the Stokes behaviors. In order to avoid numerical inaccuracies we use a very fine representation along the thickness direction and for circumventing the exponential increase in the number of degrees of freedom that such a fine representation would imply when extended to the whole laminate domain, we consider again the in-plane-out-of-plane separated representation previously introduced.

The Brinkman model is defined by:

$$\nabla p = \mu \cdot \mathbf{K}^{-1} \cdot \mathbf{v} + \eta \cdot \Delta \mathbf{v}, \tag{6.28}$$

where μ is the dynamic viscosity, $\mathbf{K}$ the layer permeability tensor and η the dynamic effective viscosity.

In the zones where Stokes model applies (resin layers) we assign a very large isotropic permeability $K = 1$ (units in the metric system) whereas in the ones occupied by the reinforcement, the permeability is assumed anisotropic, being several orders of magnitude lower, typically 10^{-8}. Thus the Darcy's component in Eq. (6.28) does not perturb the Stokes flow in the resin layers, and it becomes dominant in the reinforcement layers. Additionally by choosing this outstanding difference in permeability, representative of the one observed in liquid Resin Infusion process when highly porous distribution media are used, we also want to give the evidence that this type of problem can be addressed by the proposed approach.

6.2.2 On the Approximation Continuity

All the just addressed models imply second-order derivatives in the space coordinates, and then, after integrating by parts to recover their associated weak forms, only continuous approximations are required for both the trial and the test functions.

Thus in general in our numerical applications we considered standard piecewise linear functions for approximating the 2D fields defined in Ω and the 1D defined in $\mathscr{I}$. Obviously higher order approximations are possible with the only constraint of ensuring continuity.

Electromagnetic models derived from Maxwell's equations usually involve vector potentials that appears in the weak form affected by the *curl* operator. Let $\mathbf{A} = (A_x, A_y, A_z)$ be such a potential. We have

$$(\nabla \times \mathbf{A})^T = \left(\frac{\partial A_z}{\partial y} - \frac{\partial A_y}{\partial z}, \frac{\partial A_x}{\partial z} - \frac{\partial A_z}{\partial x}, \frac{\partial A_y}{\partial x} - \frac{\partial A_x}{\partial y} \right), \tag{6.29}$$

which proves that A_x must be continuous with respect to the coordinates y and z but should be preferably discontinuous in the x coordinate in order to ensure the transfer conditions from one medium to other. The same reasoning applies for the other components of the vector potential.

If for the sake of clarity we consider in what follows the 2D case defined by the coordinates (x, z) we could approximate A_x and A_z from

$$\begin{cases} A_x \approx \sum\limits_{i=1}^{i=N} X_i^x(x) \cdot Z_i^x(z) \\ A_z \approx \sum\limits_{i=1}^{i=N} X_i^z(x) \cdot Z_i^z(z) \end{cases}, \tag{6.30}$$

where X_i^x and Z_i^x are approximated using piecewise constant (discontinuous) and linear (continuous) functions of the x and z coordinates respectively. On the other hand X_i^z and Z_i^z are approximated using piecewise linear (continuous) and constant (discontinuous) functions of the x and z coordinates respectively. This simple choice ensures the continuity requirement just specified, and constitutes the simplest generalization of the so-called Nedelec's elements usually considered in electromagnetic numerical simulations.

6.3 Conclusions

In this chapter we revisited the use of separated representations for solving models defined in degenerated domains. The issue related to the extremely fine meshes required for capturing rich behaviors along the thickness direction was circumvented by considering in-plane-out-of-plane separated representations of the fields and model parameters involved in the models as well as the geometries in which they are defined.

In the case of quite simple behaviors plate or shell elements were introduced and widely used in many engineering applications. However, in the case of richer behaviors and/or geometries these simplified models fail to describe the more complex behaviors of the associated solutions. 3D solutions as previously discussed are too expensive from a computational point of view, however the use of in-plane-out-of-plane separated representations allows calculating 3D solutions with a computational complexity characteristic of 2D models. In the case of models involving the *curl* operator we proved that the decomposition also applies and that the approximation can be enhanced by enforcing the approximation discontinuity, by generalizing Nedelec's approximations widely considered in the finite element framework.

References

1. F. Chinesta, A. Ammar, A. Leygue, R. Keunings, An overview of the proper generalized decomposition with applications in computational rheology. J. Nonnewton. Fluid Mech. **166**, 578–592 (2011)
2. Ch. Ghnatios, F. Chinesta, Ch. Binetruy, The squeeze flow of composite laminates. Int. J. Mater. Form. (2013). doi:10.1007/s12289-013-1149-4
3. A. Leygue, F. Chinesta, M. Beringhier, T.L. Nguyen, J.C. Grandidier, F. Pasavento, B. Schrefler, Towards a framework for non-linear thermal models in shell domains. Int. J. Numer. Meth. Heat Fluid Flow **23**(1), 55–73 (2013)
4. F. Chinesta, A. Leygue, B. Bognet, Ch. Ghnatios, F. Poulhaon, F. Bordeu, A. Barasinski, A. Poitou, S. Chatel, S. Maison-Le-Poec, First steps towards an advanced simulation of composites manufacturing by automated tape placement. Int. J. Mater. Form. doi:10.1007/s12289-012-1112-9
5. B. Bognet, A. Leygue, F. Chinesta, A. Poitou, F. Bordeu, Advanced simulation of models defined in plate geometries: 3D solutions with 2D computational complexity. Comput. Methods Appl. Mech. Eng. **201**, 1–12 (2012)
6. B. Bognet, A. Leygue, F. Chinesta, On the fully 3D simulation of thermoelastic models defined in plate geometries. Eur. J. Comput. Mech **21**(1–2), 40–51 (2012)
7. B. Bognet, A. Leygue, F. Chinesta, Separated representations of 3D elastic solutions in shell geometries. Adv. Model. Simul. Eng. Sci. (2014). doi:10.1186/2213-7467-1-4
8. A. Ammar, F. Chinesta, E. Cueto, Coupling finite elements and propergeneralized decomposition. Int. J. Multiscale Comput. Eng. **9/1**, 17–33 (2011)

Chapter 7
A New Methodological Approach to Process Optimization

The best way to predict the future is to invent it.

—Alan Kay

In general, optimization implies definition of a cost function and a search for optimum process parameters (e.g the temperature of a heater, the temperature of a material approaching a die, flow rate, ... in the case of materials flowing into a heated die) defining the minimum of that cost function. The process starts by choosing a tentative set of process parameters. Then the process is simulated by discretizing the equations defining the model of the process. The solution of the model is the most costly step of the optimization procedure. As soon as that solution is available, the cost function can be evaluated and its optimality checked. If the chosen parameters do not define a minimum of the cost function, the process parameters should be updated and the solution recomputed. The procedure continues until reaching the minimum of the cost function. Obviously, nothing ensures that such a minimum is global, so more sophisticated procedures exist in order to explore the domain defined by the parameters and escape from local minimums traps. The parameter updating is carried out in a way ensuring reduction of the cost function. Many techniques update the model parameters in order to move along the cost function gradient. However, for identifying the direction of the gradient one should compute not only the fields involved in the model but also the derivatives of such fields with respect to the different process parameters. The evaluation of these derivatives is not in general an easy task. Conceptually, one could imagine that by perturbing slightly only one of the parameters involved in the process optimization and then solving the resulting model, one could estimate, using a finite difference formula, the derivative of the cost function with respect to the perturbed parameter. By perturbing sequentially all the parameters we could have access to the derivatives of the cost function with respect to all the process parameters (that is, the sensibilities) that define the cost function gradient, on which the new trial set of parameters should be chosen. There are more efficient strategies for calculating sensibilities and also techniques that do not require such sensibilities to carried out optimization as the ones based on genetic

F. Chinesta and E. Cueto, *PGD-Based Modeling of Materials, Structures and Processes*, ESAFORM Bookseries on Material Forming, DOI: 10.1007/978-3-319-06182-5_7,

algorithms or neural networks. There are many strategies for updating the set of process parameters and the interested reader can find most of them in the books focusing on optimization procedures.

Our interest here is not a discussion on particular optimization strategies, but pointing out that standard optimization strategies need numerous direct solutions of the problem that represents the process, one solution for each tentative choice of the process parameters. The solution of such models is a tricky task that demands important computational resources and usually implies extremely large computing times.

In this chapter we describe a radically different approach, to the authors' knowledge never before explored. The approach here proposed considers the unknown process parameters as new coordinates of the model. In fact, coordinates, or space dimensions, represent the (non-necessarily physical) locations at which the solution is to be represented. Thus, strictly speaking, one could compute the solution of the problem for any value of the unknown parameters (in a bounded interval). This converts those unknown parameters into new dimensions of the space in which the model is defined. This idea seems exciting but it involves a major difficulty.

This strategy faces a challenging problem if the number of parameters of the model increases. It is well known that the number of degrees of freedom for a mesh-based discretization technique (say, finite element, finite difference, …) increases exponentially with the number of dimensions. This exponential increase of the number of degrees of freedom can be literally out of reach for nowadays computers even if the number of dimensions increases only moderately. This phenomenon is known as *curse of dimensionality*.

Of course, to efficiently deal with this problem a strategy different of mesh-based discretization methods should be employed. Here, we consider the use of Proper Generalized Decompositions (PGD). PGD techniques construct an approximation of the solution by means of a sequence of products of separable functions. These functions are determined "on the fly", as the method proceeds, with no initial assumption on their form.

Thus, the solution is computed only once and it allows access to the unknown field, as well as to the explicit expression of its derivatives, for any possible choice of the model parameters by a simple particularization of the parametric solution, that is, by a simple postprocessing.

As can be readily noticed, the potential of the technique for inverse identification, optimization, etc. seems to be huge. In what follows we propose a methodological approach in this direction. A thermal model is considered of a material moving through a die equipped with some heating devices on the die walls. We could consider as process parameters the temperatures prescribed in the different heaters, flow rate, temperature of the material coming into the die, etc. For the sake of simplicity in what follows we are restricting the parametric space to heating devices temperatures. The choice of cost function depends on each particular process. There are many choices and because here we are more interested in proposing and illustrating a new methodological modeling and optimization approach than in analyzing deeply a particular process, we will restrict our analysis to a simplified model of pultrusion

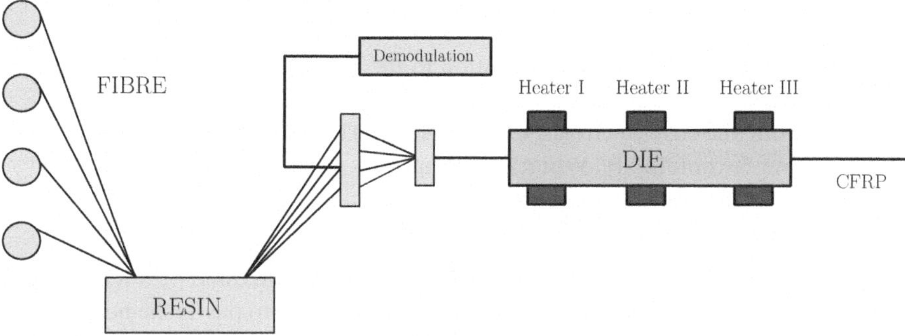

Fig. 7.1 Pultrusion process

processes. Pultrusion is a continuous process to produce constant cross-sectional profile composites. During this process, fiber reinforcements are saturated with resin, which are then pulled through a heated die. The resin gradually cures inside the die while generating heat. At the exit, pullers draw the composite out and a traveling cut-off saw cuts it at the designed length. We consider the process sketched in Fig. 7.1.

Standard optimization approaches and others based on the use of genetic algorithms have been recently proposed and applied, however the efficiency of those approaches seems to be limited to a reduced number of process parameters because one must solve a thermal model for each choice of the process parameters, and it is well known that when the dimension of the parametric space increases, the exploration of the space defined by the process parameters becomes more and more arduous, needing for numerous, sometimes excessive, solutions of the model governing the process.

If we consider the thermal model related to the pultrusion process sketched in Fig. 7.1, whose parametric space reduces to the temperatures prescribed at the three heating devices, θ_1, θ_2 and θ_3, we could summarize traditional optimization procedures as follows:

- *Until reaching a minimum of the cost function $\mathscr{C}(\theta_1, \theta_2, \theta_3)$ proceed by*:

1. Computing the temperature field related to the trial choice of the process parameters, i.e. $u(\mathbf{x}; \theta_1, \theta_2, \theta_3)$.
2. Computing the cost function $\mathscr{C}(\theta_1, \theta_2, \theta_3)$ from the just calculated thermal field.
3. Checking the optimality. While the optimum is not reached, update the process parameters by using an appropriate strategy and comeback to step 1 for another solution of the thermal model for the process parameters just updated.

In the approach that we propose in this work the procedure is substantially different. It proceeds as follows:

- Compute the thermal field for any possible choice of the process parameters: $u(\mathbf{x}, \theta_1, \theta_2, \theta_3)$ (here the heaters' temperatures play the same role as that of space coordinates), the problem becoming multidimensional.

- *Until reaching a minimum of the cost function* $\mathscr{C}(\theta_1, \theta_2, \theta_3)$ *proceed by*:

1. Particularizing the parametric solution to the considered values of the process parameters.
2. Computing the cost function $\mathscr{C}(\theta_1, \theta_2, \theta_3)$ from the just calculated thermal field.
3. Checking the optimality. While the optimum is not reached, update the process parameters by using an appropriate strategy and comeback to step 1 for another particularization of the parametric solution.

Thus, in our proposal the thermal model is solved only once and then it is particularized for any choice of the process parameters. The price to pay is the necessity of solving a multidimensional thermal model that now has as coordinates the physical space $\mathbf{x}$ and all the process parameters, i.e. the three heaters' temperatures in the example addressed here.

Obviously, the solution of the resulting multidimensional model is a tricky task if one consider a standard mesh based discretization strategy because the number of degrees of freedom increases exponentially with the dimensionality of the model. To circumvent this serious difficulty, also known as *curse of dimensionality*, we consider a separated representation of the temperature field in the PGD framework, in which the temperature reads:

$$u(\mathbf{x}, \theta_1, \theta_2, \theta_3) \approx \sum_{i=1}^{i=N} F_i(\mathbf{x}) \cdot \Theta_{1i}(\theta_1) \cdot \Theta_{2i}(\theta_2) \cdot \Theta_{3i}(\theta_3). \tag{7.1}$$

To build up such separated representation we only need to compute the functions defined in the space domain $F_i(\mathbf{x})$ and the one-dimensional functions $\Theta_{ji}(\theta_j)$, $j = 1, 2, 3$, defined in the intervals in which the heaters' temperatures can evolve. This construction will be carried out within the PGD framework carefully described and used previously. In what follows we describe the introduction of boundary conditions as extra-coordinates within the PGD formalism.

7.1 Parametric Boundary Conditions

Very often, in the optimization of an industrial process, it is necessary to solve the problem for different boundary conditions. Boundary conditions do not behave as any other parameter in the PGD, and therefore deserve some additional comments. In general, it is needed to perform a change of variable to introduce the boundary condition into the differential equation and then define it as an extra coordinate. To illustrate this procedure we consider, for the sake of simplicity and without any loss of generality, the following simple problem:

$$-\Delta u = f \text{ in } \Omega, \tag{7.2}$$

subjected to the boundary conditions:

$$u = g \neq 0 \text{ on } \Gamma \equiv \partial\Omega. \tag{7.3}$$

Let us assume, as described in [1], that we are able to find a function ψ, continuous in the closure of Ω, $\overline{\Omega}$, such that $-\Delta\psi \in L_2(\Omega)$ verifying Eq. (7.3). Then, the solution of the problem given by Eqs. (7.2), (7.3) can be obtained straightforwardly by

$$u = \psi + z, \tag{7.4}$$

where we thus face a problem in the z variable

$$-\Delta z = f + \Delta\psi \text{ in } \Omega \tag{7.5}$$

$$z = 0 \text{ on } \Gamma \tag{7.6}$$

easily solvable by the PGD method presented before.

The introduction of the value of the boundary condition (g in Eq. (7.3)) as an extra-coordinate generates an extra dimension in the problem, which means dramatically increasing the computation cost of the problem solution in classical mesh-based numerical methods for discretizing partial differential equations.

7.2 Parametric Thermal Model of a Heated Die in Pultrusion Processes

In modeling the pultrusion process, as sketched in Fig. 7.1, we consider the thermal process within the die as modeled by the following two-dimensional convection-diffusion equation:

$$\rho \cdot C\left(\frac{\partial u}{\partial t} + v\frac{\partial u}{\partial x}\right) = k\Delta u + q, \tag{7.7}$$

where k is the thermal conductivity, q is the internal heat generated by the resin curing reaction, ρ is the density, C is the specific heat and v is the extruded profile speed.

The die is equipped with three heaters as depicted in Fig. 7.2 whose temperatures constitute the process parameters to be optimized. For the sake of simplicity we consider constant profile velocity v and inlet temperature U_0, all of them assumed known. The curing kinetics was coupled with the thermal field as described in [2].

Thus, the temperature field u depends on five different coordinates, the two space coordinates (x, y) and the three temperatures prescribed in three regions on the die wall. In fact $u = u(x, y, \theta_1, \theta_2, \theta_3)$, where θ_1, θ_2 and θ_3 are the temperature of the heaters I, II and III, depicted in Fig. 7.2 respectively. The separated representation of u reads:

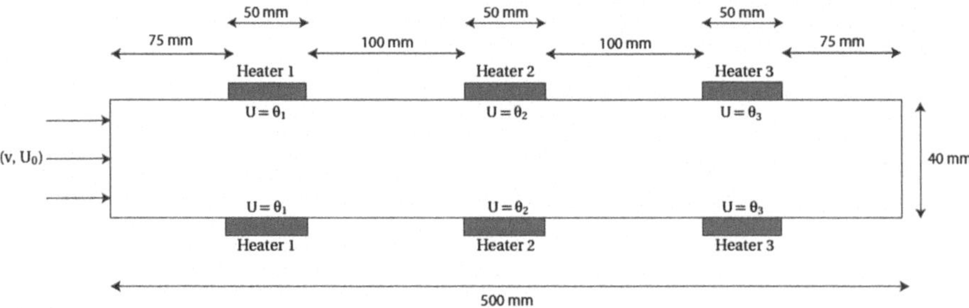

Fig. 7.2 Domain geometry and boundary conditions

$$u(x, y, \theta_1, \theta_2, \theta_3) \approx \sum_{i=1}^{i=N} F_i(x, y) \cdot \Theta_{1i}(\theta_1) \cdot \Theta_{2i}(\theta_2) \cdot \Theta_{3i}(\theta_3). \tag{7.8}$$

Temperatures are assumed prescribed as Dirichlet boundary conditions on Γ_D whereas a null flux is prescribed on the complementary part $\Gamma_N = \Gamma - \Gamma_D$, i.e. $\nabla u \cdot \mathbf{n}|_{\Gamma_N} = 0$.

The prescribed temperatures are written:

$$\begin{cases} u(x = 0, y, \theta_1, \theta_2, \theta_3) = U_0 \\ u(x \in L_1, y = 0 \text{ or } y = h, \theta_1, \theta_2, \theta_3) = \theta_1 \\ u(x \in L_2, y = 0 \text{ or } y = h, \theta_1, \theta_2, \theta_3) = \theta_2 \\ u(x \in L_3, y = 0 \text{ or } y = h, \theta_1, \theta_2, \theta_3) = \theta_3 \end{cases},$$

where L_1, L_2 and L_3 are the intervals of the x-coordinate related to heaters I, II and III respectively. $L = 0.5\,m$ is the length of the die and $h = 0.04\,m$ its width. The parameters θ_1, θ_2 and θ_3 take values in the intervals $\mathcal{I}_1$, $\mathcal{I}_2$ and $\mathcal{I}_3$ respectively.

To solve this problem, we need to define some functional products in order to introduce the boundary conditions (7.2) into the partial differential equation (four modes suffices for this end) and then proceed by applying the separated representation constructor.

7.3 Optimization Strategy

In this section we consider the parametric solution already computed:

$$u(x, y, \theta_1, \theta_2, \theta_3) \approx \sum_{i=1}^{i=N} F_i(x, y) \cdot \Theta_{1i}(\theta_1) \cdot \Theta_{2i}(\theta_2) \cdot \Theta_{3i}(\theta_3). \tag{7.9}$$

The objective of the optimization procedure consists of the determination of the process parameters θ_1, θ_2 and θ_3 in order to minimize an appropriate cost function depending on the considered physics. In this work we are only interested in describing the main ingredients involved in the proposed optimization strategy, and for this reason, in what follows, we consider a cost function without any particular physical significance.

The cost function is defined from the difference between the temperature on the outflow boundary $x = L$, $u(x = L, y; \theta_1, \theta_2, \theta_3)$, and the desired one $\overline{u}(y)$. Thus, the cost function reads:

$$\mathscr{C}(\theta_1, \theta_2, \theta_3) = \frac{1}{2} \int_0^h (u(x = L, y; \theta_1; \theta_2, \theta_3) - \overline{u}(y))^2 \, dy. \tag{7.10}$$

We look for a uniform profile of temperatures on the outlet. This kind of condition is quite difficult to fulfill due to the heat conduction mechanism as well as the internal heat generation. We consider this cost function only to prove the robustness of the proposed optimization approach.

The parametric domain is defined by $\mathscr{I} = \mathscr{I}_1 \times \mathscr{I}_2 \times \mathscr{I}_3$ in which the optimal solution is sought. The optimal solution constitutes a minimum of the cost function (7.10). We denote points in the parametric domain by θ with components $(\theta_1, \theta_2, \theta_3)$.

The algorithm starts by considering an arbitrary point within $\mathscr{I}$. Then, the gradient and Hessian defined by the cost function are computed in order to apply a Newton strategy. For computing both, we need to define an appropriate approximation of the cost function in the vicinity of the evaluation point. By defining a quadratic approximation we could compute the gradient and the Hessian, and because the approximation is quadratic the Newton algorithm is an appropriate choice for performing the minimization.

Let $\theta^0 \in \mathscr{I}$ be the starting point. We consider a small paralelepipedic volume $\mathscr{P}^0$ in the parametric space centered at that point in which the cost function will be approximated using a polynomial approximation of a certain order, and whose edges lengths l_1, l_2 and l_3, in the directions of the axes θ_1, θ_2 and θ_3 respectively, depend on the desired order of the approximation as well as on the mesh used for discretizing the intervals $\mathscr{I}_1$, $\mathscr{I}_2$ and $\mathscr{I}_3$ in which θ_1, θ_2 and θ_3, respectively, are defined. Then, a number of points θ_i^0, $i = 1, \ldots, n_{LH}$, are considered within $\mathscr{P}^0$ according to the LATIN hypercube technique.

The cost function $\mathscr{C}_i^0$, $i = 1, \ldots, n_{LH}$ is computed at those points θ_i^0, $i = 1, \ldots, n_{LH}$. Now, a quadratic approximation of $\mathscr{C}(\theta_1, \theta_2, \theta_3)$ could be defined in $\mathscr{P}^0$, $\mathscr{C}^h(\theta \in \mathscr{P}^0)$, and then both the gradient and the Hessian computed.

Obtaining a smooth enough, quadratic approximation of a field given by scattered data is not always easy. For this purpose we make use of a centered moving least square (MLS).

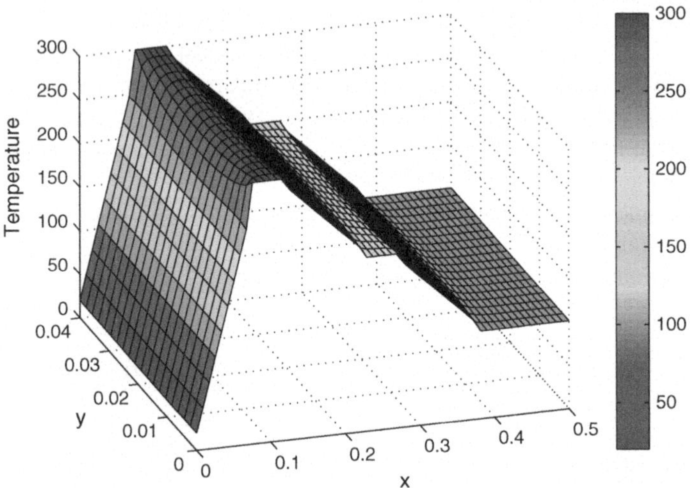

Fig. 7.3 Multidimensional solution particularized for the heaters temperatures $\theta_1 = 300\,^\circ$C, $\theta_2 = 200\,^\circ$C and $\theta_3 = 100\,^\circ$C

7.3.1 Numerical Example

In order to illustrate the above presented procedure, we consider the geometry shown in Fig. 7.2 and the following material and process parameters: $v = 0.26$, $\rho = 1560$, $C = 1700$ and $k = 3.7 \cdot 10^{-7}$, all these values expressed in the metric system.

First, proceeding as indicated in the previous sections, we compute the solution for all possible boundary conditions:

$$u(x, y, \theta_1, \theta_2, \theta_3) \approx \sum_{i=1}^{i=N} F_i(x, y) \cdot \Theta_{1i}(\theta_1) \cdot \Theta_{2i}(\theta_2) \cdot \Theta_{3i}(\theta_3). \tag{7.11}$$

This parametric solution can then be particularized for any choice of the temperatures of the heaters. One of the possible solutions related to a particular choice of the heaters temperatures is depicted in Fig. 7.3.

If we compare the solution represented in Fig. 7.3 to the one obtained by the finite element method for the same values of the heaters' temperatures, the difference (still using the L_2-norm) is found to be lower than 10^{-5} for N as little as 30 modes.

Now, we are considering the optimization process with respect the cost function given by Eq. (7.10), where the target temperature at the outflow is fixed to $\bar{u} = 150\,^\circ$C.

The temperature distribution in the whole die for these optimized values is shown in Fig. 7.4, for the optimal temperatures of the three heaters:

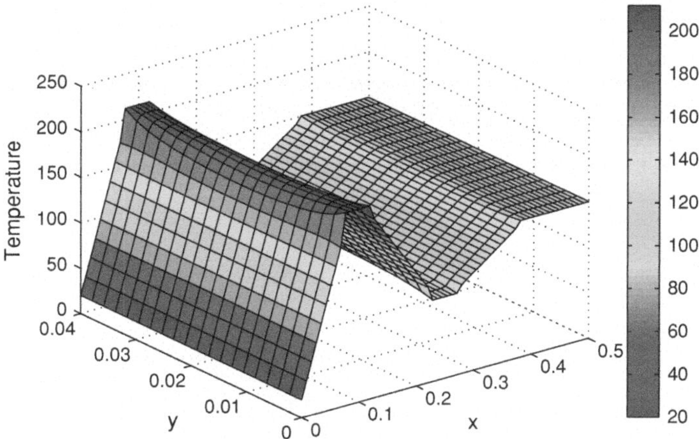

Fig. 7.4 Final, optimized distribution of temperatures

$$\begin{aligned} \theta_1^{\mathrm{opt}} &= 211.9\,^{\circ}\mathrm{C}, \\ \theta_2^{\mathrm{opt}} &= 89.3\,^{\circ}\mathrm{C}, \\ \theta_3^{\mathrm{opt}} &= 150.2\,^{\circ}\mathrm{C}. \end{aligned} \tag{7.12}$$

The value of the associated cost function when the three optimal temperatures were considered was of $3.1 \cdot 10^{-7}$.

Remark 7.3.1 Note that the values taken in this academic example do not correspond to any physically meaningful ones. The obtained temperatures could have an impact on the resin viscosity and therefore in its thermal conductivity. We are here mostly interested in introducing the optimization strategy.

7.4 Conclusion

The technique here proposed could allow impressive computing efficiencies, however it is too early for quantifying the reachable computing time savings. The present technique only needs one solution of a multidimensional model, that thanks to the PGD features can be performed easily and very fast. As soon as this parametric solution is known we can imagine a panoply of optimization problems, involving different cost functions and numerical strategies, performed on-line without the necessity of solving the model again. The optimization described above was carried out by using Matlab and a standard laptop in 0.0015 s.

References

1. D. Gonzalez, A. Ammar, F. Chinesta, E. Cueto, Recent advances in the use of separated representations. Int. J. Numer. Meth. Eng **81**(5), 637–659 (2010)
2. E. Pruliere, J. Ferec, F. Chinesta, A. Ammar, An efficient reduced simulation of residual stresses in composites forming processes. Int. J. Mater. Form. **3**(2), 1339–1350 (2010)

Chapter 8
A New Methodological Approach for Shape Optimization

In this chapter we focus on shape optimization that involves the appropriate choice of some parameters defining the problem geometry. The main objective of this work is to describe an original approach for computing an off-line parametric solution. That is, a solution able to include information for different parameter values and also allowing us to compute readily the sensitivities. The curse of dimensionality is circumvented by invoking the Proper Generalized Decomposition (PGD) introduced in former works, which is applied here to compute geometrically parametrized solutions.

8.1 Introduction

In general, shape optimization implies a definition of a cost function and a search forthe optimum parameters (e.g. geometrical parameters describing the family of possible shapes) defining the minimum of that cost function. The process starts by choosing a tentative set of parameters. Then the process is simulated by discretizing the equations defining the problem in the geometry defined by the tentative geometrical parameters. This solution is the most costly step of the optimization procedure. As soon as that solution is available, the cost function can be evaluated and its optimality checked. If the chosen parameters do not define a minimum (at least local) of the cost function, the process parameters should be updated and the solution recomputed. The procedure continues until reaching the minimum of that cost function. The general framework was comprehensively described in the previous chapter.

This chapter represents a step forward by introducing the geometrical parameters as extra-coordinates. We will prove here that this extension is far from trivial.

If we consider a simple model, e.g. the steady heat equation defined in the domain sketched in Fig. 8.1, whose parametric space reduces to the horizontal and vertical displacement of the upper right corner, θ_1 and θ_2, we could summarize traditional optimization procedures as follows:

F. Chinesta and E. Cueto, *PGD-Based Modeling of Materials, Structures and Processes*, ESAFORM Bookseries on Material Forming, DOI: 10.1007/978-3-319-06182-5_8,

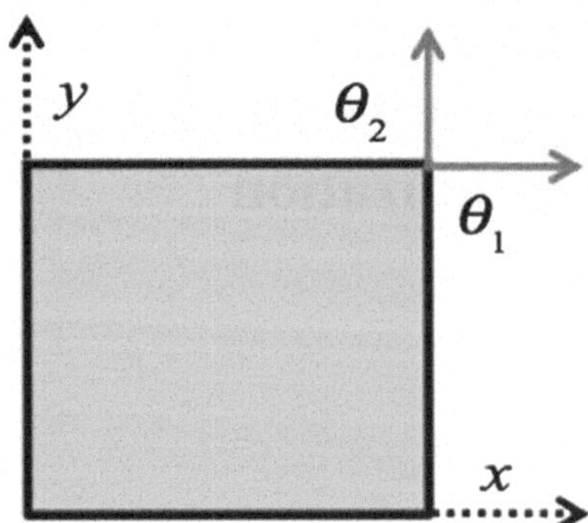

Fig. 8.1 Parametric domain

- *Until reaching a minimum of the cost function $\mathscr{C}(\theta_1, \theta_2)$ proceed by:*

  1. Computing the unknown field related to the trial choice of the geometry, i.e., $u(\mathbf{x}; \theta_1, \theta_2)$.
  2. Computing the cost function $\mathscr{C}(\theta_1, \theta_2)$ from the just calculated thermal field.
  3. Checking the optimality. While the optimum is not reached, update the geometry, i.e., θ_1 and θ_2, by using an appropriate strategy and come back to step 1 and solve again the model in the just updated geometry.

In the approach that we propose in this work the procedure is substantially different. It proceeds as follows:

- Compute the unknown field in any possible geometry: $u(\mathbf{x}, \theta_1, \theta_2)$ (here the location of the upper-right corner plays the same role as the space coordinates), the problem becoming multidimensional.
- *Until reaching a minimum of the cost function $\mathscr{C}(\theta_1, \theta_2)$ proceed by:*

  1. Particularizing the parametric solution to the considered values of the geometrical parameters.
  2. Computing the cost function $\mathscr{C}(\theta_1, \theta_2)$ from $u(\mathbf{x}, \theta_1, \theta_2)$.
  3. Checking the optimality. While the optimum is not reached, update the geometry by using an appropriate strategy and come back to step 1 and particularize again the parametric solution.

Thus, in our proposal the model is solved only once and then it is particularized for any choice of the geometry. The price to pay is the necessity of solving a multidimensional thermal model that now has as coordinates the physical space $\mathbf{x}$ and all the geometrical parameters, in the present example the two extra-coordinates θ_1 and θ_2 defining the location of the upper-right corner.

Obviously, the solution of the resulting multidimensional model is a tricky task if one considers a standard mesh based discretization strategy because the number of degrees of freedom increases exponentially with the dimensionality of the model. To circumvent this serious difficulty, also known as *curse of dimensionality*, we consider a separated representation of the unknown field within the PGD framework that reads:

$$u(\mathbf{x}, \theta_1, \theta_2) \approx \sum_{i=1}^{i=N} F_i(\mathbf{x}) \cdot \Theta_{1i}(\theta_1) \cdot \Theta_{2i}(\theta_2) \tag{8.1}$$

To build up such separated representation we only need to compute the functions defined in the space domain $F_i(\mathbf{x})$ and the one-dimensional functions $\Theta_{ji}(\theta_j)$, $j = 1, 2$, defined in the intervals in which the location of the upper-right corner can evolve. For this purpose we consider the standard PGD procedure, but prior tousing it we address the introduction of geometrical parameters as extra-coordinates.

8.2 Introducing Geometrical Parameters as Extra-Coordinates

In this section, which constitutes the heart of this work, we consider for the sake of simplicity the 2D steady state heat equation

$$\Delta u(x, y) = -f, \tag{8.2}$$

with a constant source term, i.e., $f = cte$, and homogenous boundary conditions, defined in the domain Ω^θ that results when the unit square $\Omega^r =]0, 1[^2$ is distorted by moving its upper-right corner θ_1 and θ_2 in the horizontal and vertical directions respectively.

The weak form of Eq. (8.2) reads

$$\int_{\Omega^\theta} \nabla u^* \cdot \nabla u \, d\Omega = \int_{\Omega^\theta} u^* \cdot f \, d\Omega. \tag{8.3}$$

We are now interested in introducing both displacements θ_1 and θ_2 as extra-coordinates and then computing the general parametric solution $u(x, y, \theta_1, \theta_2)$.

8.2.1 A First Tentative Geometrical Transformation

In order to explicit the dependence of the model on both geometrical parameters we consider a first tentative geometrical transformation $(X, Y) \in \Omega^r \rightarrow (x, y) \in \Omega^\theta$:

$$\begin{cases} x = X + X \cdot Y \cdot \theta_1 \\ y = Y + X \cdot Y \cdot \theta_2 \end{cases}, \tag{8.4}$$

that defines the Jacobian $\mathbf{J}$

$$\mathbf{J} = \begin{pmatrix} \frac{\partial x}{\partial X} & \frac{\partial y}{\partial X} \\ \frac{\partial x}{\partial Y} & \frac{\partial y}{\partial Y} \end{pmatrix} = \begin{pmatrix} 1 + Y \cdot \theta_1 & Y \cdot \theta_2 \\ X \cdot \theta_1 & 1 + X \cdot \theta_2 \end{pmatrix}. \tag{8.5}$$

The inverse transformation that we will denote by $\mathbf{Q}$, can be easily computed from the inverse of the Jacobian, i.e., $\mathbf{Q} = \mathbf{J}^{-1}$:

$$\mathbf{Q} = \begin{pmatrix} \frac{\partial X}{\partial x} & \frac{\partial Y}{\partial x} \\ \frac{\partial X}{\partial y} & \frac{\partial Y}{\partial y} \end{pmatrix} = \frac{1}{\det\mathbf{J}} \begin{pmatrix} 1 + X \cdot \theta_2 & -Y \cdot \theta_2 \\ -X \cdot \theta_1 & 1 + Y \cdot \theta_1 \end{pmatrix}. \tag{8.6}$$

With this transformation the weak form (8.3) reads:

$$\int_{\Omega^r} \left(\mathbf{Q} \cdot \nabla^r u^*\right) \cdot \left(\mathbf{Q} \cdot \nabla^r u\right) \cdot \det\mathbf{J} \; d\Omega = \int_{\Omega^r} u^* \cdot f \cdot \det\mathbf{J} \, d\Omega, \tag{8.7}$$

where $\nabla^r = (\partial/\partial X, \partial/\partial Y)$.

We can notice that the left-hand member involves terms that can be expressed in a separated form, implying a finite sum of products of functions of $\mathbf{X} = (X, Y)$, functions of θ_1 and functions of θ_2. However we can notice the appearance of a $\det\mathbf{J}$ in the denominator, and this is finally the only tricky point because: (i) the expression of the inverse of a separated function in a separated form can involve too many terms, and (ii) in order to build up such separated representation

$$\frac{1}{\det\mathbf{J}} \approx \sum_{i=1}^{P} J_i^x(\mathbf{X}) \cdot J_i^1(\theta_1) \cdot J_i^2(\theta_2), \tag{8.8}$$

we should apply a HOSVD (high order singular value decomposition) and when we operate in a multidimensional space involving many extra-coordinates, its application becomes delicate and in any case non optimal.

8.2.2 Looking for Simpler Transformations

In order to avoid the difficulties related to the necessity of representing the inverse of $\det\mathbf{J}$ in a separated form, we look for alternative transformations, the optimal choice being those resulting in a separated expression of the inverse of $\det\mathbf{J}$.

For this purpose imagine the quadrilateral domain Ω^θ depicted in Fig. 8.1, obtained by distorting the unit square, defined by the four corners $\mathscr{P}_1 = (0, 0)$, $\mathscr{P}_2 = (1, 0)$, $\mathscr{P}_3 = (1 + \theta_1, 1 + \theta_2)$ and $\mathscr{P}_4 = (0, 1)$, decomposed into two triangles $\mathscr{T}_1$ and $\mathscr{T}_2$, the first one defined by $(\mathscr{P}_1, \mathscr{P}_3, \mathscr{P}_4)$ and the second one by $(\mathscr{P}_1, \mathscr{P}_2, \mathscr{P}_3)$.

We consider the reference triangle $\mathscr{T}^r$ whose corners are given by $\mathscr{P}_1^r = (0, 0)$, $\mathscr{P}_2^r = (1, 0)$, $\mathscr{P}_3^r = (0, 1)$ and the two geometrical transformations $\mathscr{T}_1 \rightarrow \mathscr{T}^r$ and $\mathscr{T}_2 \rightarrow \mathscr{T}^r$.

The interpolation of any function $u(X, Y)$ defined in $\mathscr{T}^r$ can be written as:

$$u(\mathbf{X} \in \mathscr{T}^r) = U_1 \cdot N^1(\mathbf{X}) + U_2 \cdot N_2(\mathbf{X}) + U_3 \cdot N_3(\mathbf{X}), \tag{8.9}$$

where U_i, $i = 1, 2, 3$, denotes the value of the field u at positions $\mathscr{P}_1^r$, $\mathscr{P}_2^r$ and $\mathscr{P}_3^r$ respectively and the shape functions $N_i(\mathbf{X})$ verify the Kroenecker's delta property and read:

$$\begin{cases} N_1(X, Y) = 1 - X - Y \\ N_2(X, Y) = X \\ N_3(X, Y) = Y \end{cases}. \tag{8.10}$$

When applying this interpolation to the coordinate transformation we obtain in $\mathscr{T}_1$:

$$\begin{cases} x(\mathbf{X}) = x_1 \cdot N_1(\mathbf{X}) + x_3 \cdot N_2(\mathbf{X}) + x_4 \cdot N_3(\mathbf{X}) \\ y(\mathbf{X}) = y_1 \cdot N_1(\mathbf{X}) + y_3 \cdot N_2(\mathbf{X}) + y_4 \cdot N_3(\mathbf{X}) \end{cases}, \tag{8.11}$$

and a similar expression in $\mathscr{T}_2$:

$$\begin{cases} x(\mathbf{X}) = x_1 \cdot N_1(\mathbf{X}) + x_2 \cdot N_2(\mathbf{X}) + x_3 \cdot N_3(\mathbf{X}) \\ y(\mathbf{X}) = y_1 \cdot N_1(\mathbf{X}) + y_2 \cdot N_2(\mathbf{X}) + y_3 \cdot N_3(\mathbf{X}) \end{cases}, \tag{8.12}$$

where $\mathbf{x}_i = (x_i, y_i)$ denotes the coordinates of points $\mathscr{P}_i$, $i = 1, \ldots, 4$. In the case here considered we get:

$$\begin{cases} \mathbf{x}_1 = (0, 0) \\ \mathbf{x}_2 = (1, 0) \\ \mathbf{x}_3 = (1 + \theta_1, 1 + \theta_2) \\ \mathbf{x}_4 = (0, 1) \end{cases}. \tag{8.13}$$

Thus, the Jacobian related to the transformation $\mathscr{T}^r \rightarrow \mathscr{T}_1$, $\mathbf{J}_1$, becomes independent of the space coordinates because of the linearity of the transformation, and in the present case reads:

$$\mathbf{J}_1 = \begin{pmatrix} 1 + \theta_1 & 1 + \theta_2 \\ 0 & 1 \end{pmatrix}, \tag{8.14}$$

and analogously for the transformation $\mathscr{T}^r \rightarrow \mathscr{T}_2$ whose Jacobian $\mathbf{J}_2$ reads

$$\mathbf{J}_2 = \begin{pmatrix} 1 & 0 \\ 1 + \theta_1 & 1 + \theta_2 \end{pmatrix}. \tag{8.15}$$

In this particular case $\det \mathbf{J}_1 = 1 + \theta_1$ and $\det \mathbf{J}_2 = 1 + \theta_2$, and their inverse can be written from a single functional product.

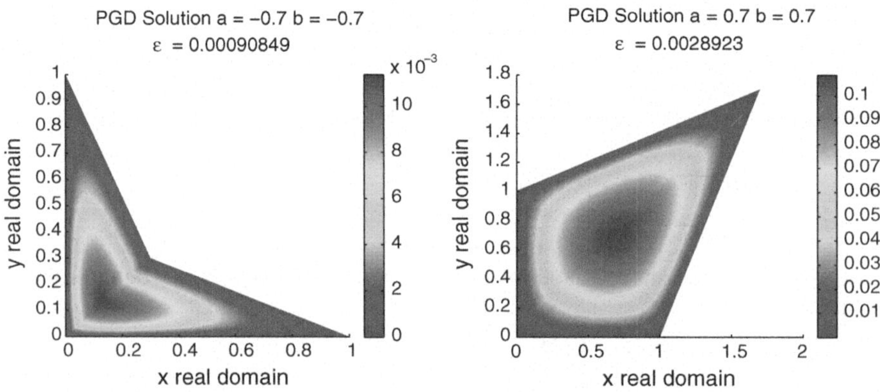

Fig. 8.2 Solution of the thermal model for: (*top*) $(\theta_1, \theta_2) = (-0.7, -0.7)$ and (*bottom*) $(\theta_1, \theta_2) = (0.7, 0.7)$

8.2.3 Numerical Example

We can now compute the model solution in the separated form

$$u(\mathbf{x}, \theta_1, \theta_2) \approx \sum_{i=1}^{i=N} F_i(\mathbf{x}) \cdot \Theta_{1i}(\theta_1) \cdot \Theta_{2i}(\theta_2), \tag{8.16}$$

by applying the standard PGD constructor.

In order to approximate functions $F_i(\mathbf{x})$ we consider a mesh on Ω^r composed of linear triangles such that no element crosses the interface $\Gamma = \overline{\mathscr{T}}_1 \cap \overline{\mathscr{T}}_2$. This approximation is then considered for solving the second order boundary value problem in the spatial coordinates related to those functions $F_i(\mathbf{x})$.

When assuming $(\theta_1, \theta_2) \in [-0.7, 0.7] \times [-0.7, 0.7]$ we can compute from one single execution, the solution for any geometry obtained by moving the upper-right corner of the unit square any value θ_1 and θ_2 in the horizontal and vertical directions respectively.

Figure 8.2 depicts the limit solutions obtained from $(\theta_1, \theta_2) = (-0.7, -0.7)$ and $(\theta_1, \theta_2) = (0.7, 0.7)$. Both solutions were compared with the ones computed by using finite elements in both geometries and both agreed to a high level of accuracy.

8.2.4 General Setting

In the just analyzed scenario the geometrical parameters are involved only in the coordinates of $\mathscr{P}_3$, (x_3, y_3). We are considering now a more general situation in which the geometrical transformation $\mathscr{T}^r \rightarrow \mathscr{T}$ reads:

$$\begin{cases} x(\mathbf{X}) = x_1 \cdot N_1(\mathbf{X}) + x_2 \cdot N_2(\mathbf{X}) + x_3 \cdot N_3(\mathbf{X}) \\ y(\mathbf{X}) = y_1 \cdot N_1(\mathbf{X}) + y_2 \cdot N_2(\mathbf{X}) + y_3 \cdot N_3(\mathbf{X}) \end{cases}. \tag{8.17}$$

We assume the following dependences of the coordinates $\mathbf{x}_i$ on the geometrical parameters θ_i^x and θ_i^y, $i = 1, 2, 3$:

$$\begin{cases} x_1 = f_1^x(\theta_1^x) \\ x_2 = f_2^x(\theta_2^x) \\ x_3 = f_3^x(\theta_3^x) \\ y_1 = f_1^y(\theta_1^y) \\ y_2 = f_2^y(\theta_2^y) \\ y_3 = f_3^y(\theta_3^y) \end{cases}, \tag{8.18}$$

from which we obtain the following general expression of the Jacobian

$$\mathbf{J} = \begin{pmatrix} -f_1^x(\theta_1^x) + f_2^x(\theta_2^x) & -f_1^y(\theta_1^y) + f_2^y(\theta_2^y) \\ -f_1^x(\theta_1^x) + f_3^x(\theta_3^x) & -f_1^y(\theta_1^y) + f_3^y(\theta_3^y) \end{pmatrix}. \tag{8.19}$$

The inverse of its determinant must be separated but at least it does not depend on the space coordinates and many times, simpler expressions are derived when not all the coordinates $\mathbf{x}_i$ are related to geometrical parameters.

8.3 Numerical Results

In this section we apply the technique just described in some more complex scenarios.

8.3.1 Rectangle with a Parametrized Edge

The first numerical experiment concerns a domain Ω^θ obtained by distorting the rectangular reference domain $\Omega^r =]0, 3[\times]0, 1[$. For this purpose we assume Ω^r described by eight control points $\mathscr{P}_i$, $i = 1, \ldots, 8$, with coordinates

$$\begin{cases} \mathscr{P}_1^r = (0, 0) \\ \mathscr{P}_2^r = (1, 0) \\ \mathscr{P}_3^r = (2, 0) \\ \mathscr{P}_4^r = (3, 0) \\ \mathscr{P}_5^r = (3, 1) \\ \mathscr{P}_6^r = (2, 1) \\ \mathscr{P}_7^r = (1, 1) \\ \mathscr{P}_8^r = (0, 1) \end{cases}. \tag{8.20}$$

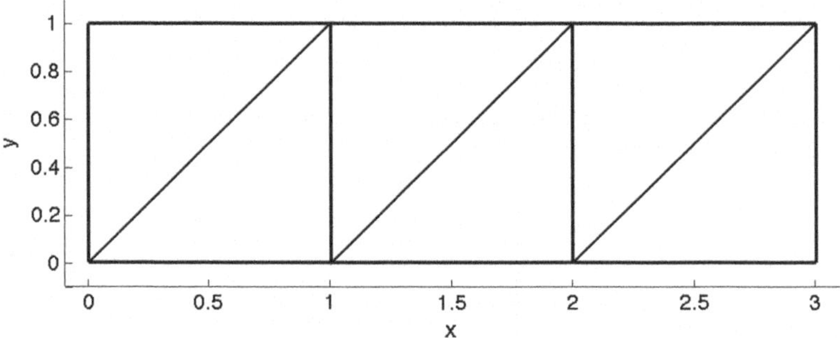

Fig. 8.3 Triangulation defining the geometrical transformation

In order to apply the procedure previously described we consider the six triangles: $\mathcal{T}_1 = (\mathcal{P}_1, \mathcal{P}_7, \mathcal{P}_8)$, $\mathcal{T}_2 = (\mathcal{P}_1, \mathcal{P}_2, \mathcal{P}_7)$, $\mathcal{T}_3 = (\mathcal{P}_2, \mathcal{P}_6, \mathcal{P}_7)$, $\mathcal{T}_4 = (\mathcal{P}_2, \mathcal{P}_3, \mathcal{P}_6)$, $\mathcal{T}_5 = (\mathcal{P}_3, \mathcal{P}_5, \mathcal{P}_6)$ and $\mathcal{T}_6 = (\mathcal{P}_3, \mathcal{P}_4, \mathcal{P}_5)$, as depicted in Fig. 8.3.

The distorted domain Ω is obtained by moving vertically points $\mathcal{P}_i^r, i = 5, \ldots, 8$, i.e.:

$$\begin{cases} \mathcal{P}_1 = (0, 0) \\ \mathcal{P}_2 = (1, 0) \\ \mathcal{P}_3 = (2, 0) \\ \mathcal{P}_4 = (3, 0) \\ \mathcal{P}_5 = (3, 1 + \theta_1) \\ \mathcal{P}_6 = (2, 1 + \theta_2) \\ \mathcal{P}_7 = (1, 1 + \theta_3) \\ \mathcal{P}_8 = (0, 1 + \theta_4) \end{cases} . \tag{8.21}$$

with $\theta_i \in [-0.3, 0.3]$, $i = 1, \ldots, 4$.

The resulting solution separated representation involves 40 terms

$$u(\mathbf{x}, \theta_1, \theta_2, \theta_3, \theta_4) \approx \sum_{i=1}^{i=40} F_i(\mathbf{x}) \cdot \Theta_{1_i}(\theta_1) \cdot \Theta_{2_i}(\theta_2) \cdot \Theta_{3_i}(\theta_3) \cdot \Theta_{4_i}(\theta_4). \tag{8.22}$$

Functions $\Theta_{j_i}(\theta_j)$ were approximated from a one-dimensional nodal distribution consisting of 13 nodes uniformly distributed in the interval $[-0.3, 0.3]$. Functions $F_i(\mathbf{x})$ were approximated by using the linear finite element approximations on the mesh depicted in Fig. 8.4.

Figure 8.5 depicts the three most significant functions $F_i(\mathbf{x})$ and Fig. 8.6 the three most significant functions $\Theta_{j_i}(\theta_j)$, $j = 1, \ldots, 4$, $i = 1, 2, 3$.

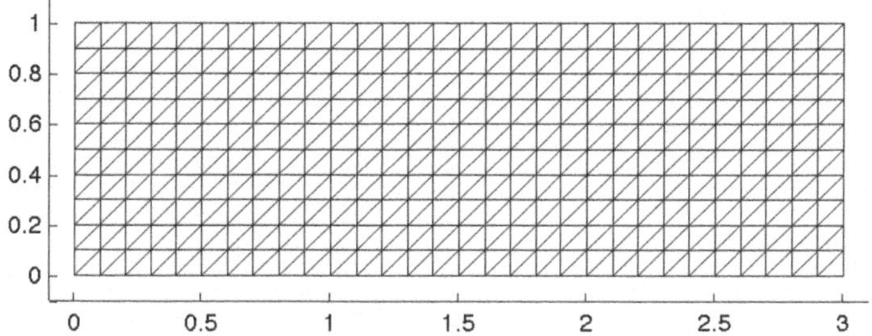

Fig. 8.4 Mesh considered for approximating functions $F_i(\mathbf{x})$

Figure 8.7 compares the finite element solution related to a domain defined by

$$\begin{cases} \theta_1 = -0.15 \\ \theta_2 = 0.3 \\ \theta_3 = -0.3 \\ \theta_4 = 0.3 \end{cases}, \tag{8.23}$$

with the one that results from the particularization of the PGD parametric solution $u(\mathbf{x}, \theta_1 = -0.15, \theta_2 = 0.3, \theta_3 = -0.3, \theta_4 = 0.3)$. The difference between the two solutions was, using an L^2 norm, of around 10^{-2}. Figure 8.8 shows the evolution or the error (defined as the difference of the PGD solution and the finite element solution considered as the reference one) as a function of the number of terms involved in the separated representation. In this same figure we represent the evolution of the computing time with the number of terms involved in the approximation.

Finally, Fig. 8.9 depicts the FE solutions and the particularized parametric PGD solutions for two other different geometries. Again, both solutions are in perfect agreement.

8.3.2 Rectangle Connected with a Circle

In this section we consider again the heat equation in a domain that results from the dilatation of a unit square that is connected through its right edge with a semicircle as depicted in Fig. 8.10.

Thus $\Omega^r = \Omega^{r1} \cup \Omega^{r2}$:

$$\begin{cases} \Omega^{r1} -]0, L = 1[\times]0, H = 1[\\ \Omega^{r2} = (\mathscr{C}((1, 0.5); R = 0.5) \cap X \geq 1) \end{cases}, \tag{8.24}$$

where $\mathscr{C}((1, 0.5); R = 0.5) \cap X \geq 1$ represents the part of the circle of radius $R = 0.5$ centered at point $(1, 0.5)$ with $X \geq 1$.

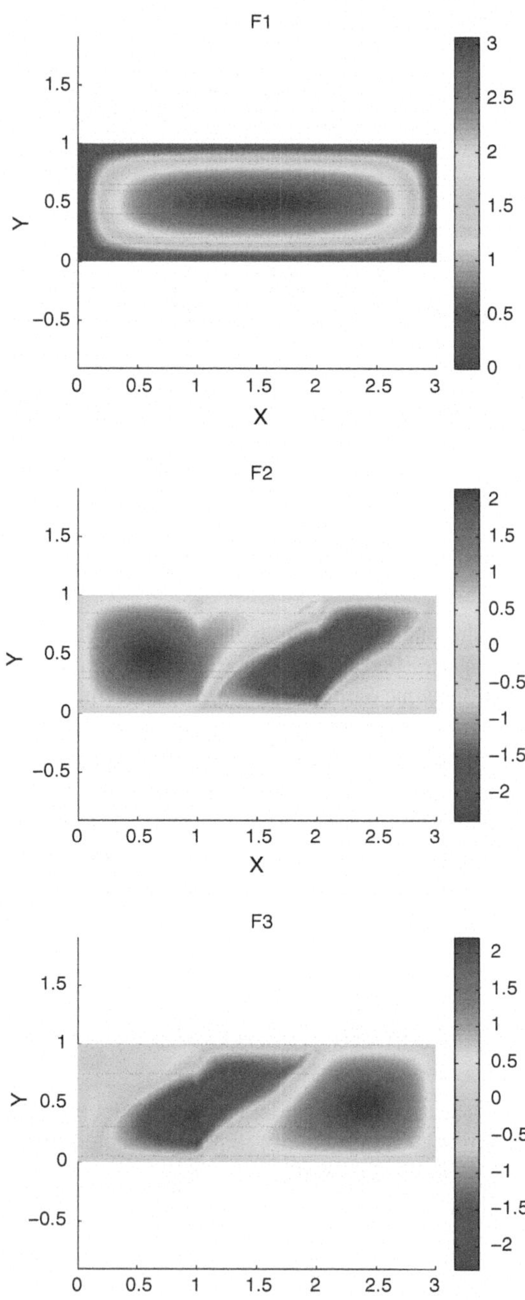

Fig. 8.5 Most significant spatial modes

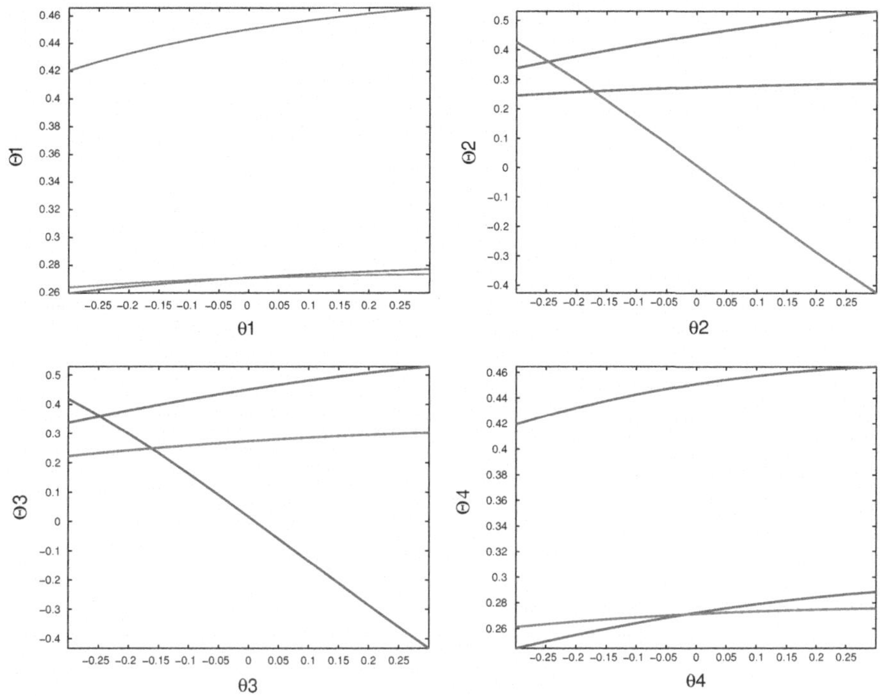

Fig. 8.6 Most significant modes $\Theta_{j_i}(\theta_j)$

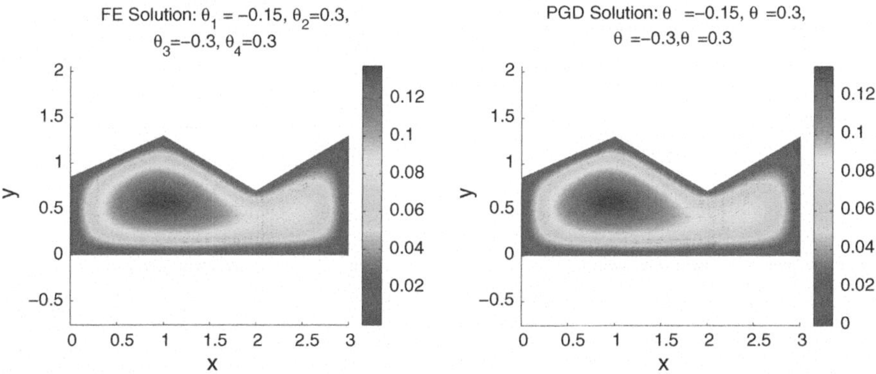

Fig. 8.7 Comparing the finite element solution (*left*) with the particularized PGD parametric solution (*right*)

We consider two geometrical parameters θ_1 and θ_2 for describing the family of geometries that we are considering: (i) the rectangle length $l = L \cdot \theta_1$ and (ii) the semicircle radius $r = R \cdot \theta_2$.

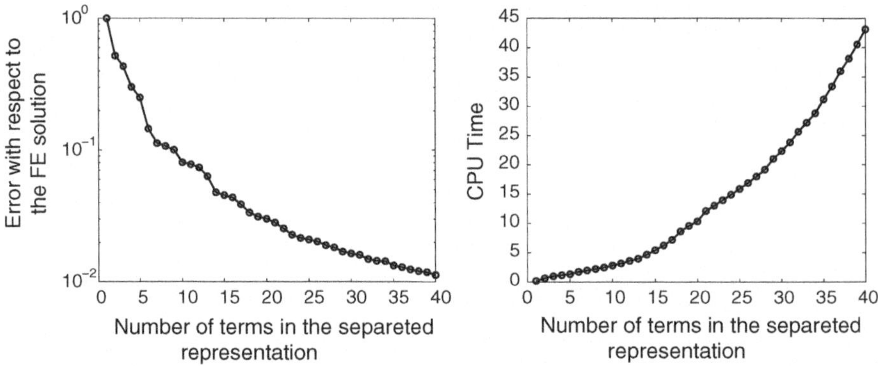

Fig. 8.8 Evolution of the solution error (*left*) and the CPU time (*right*) with the number of terms involved in the separated representation

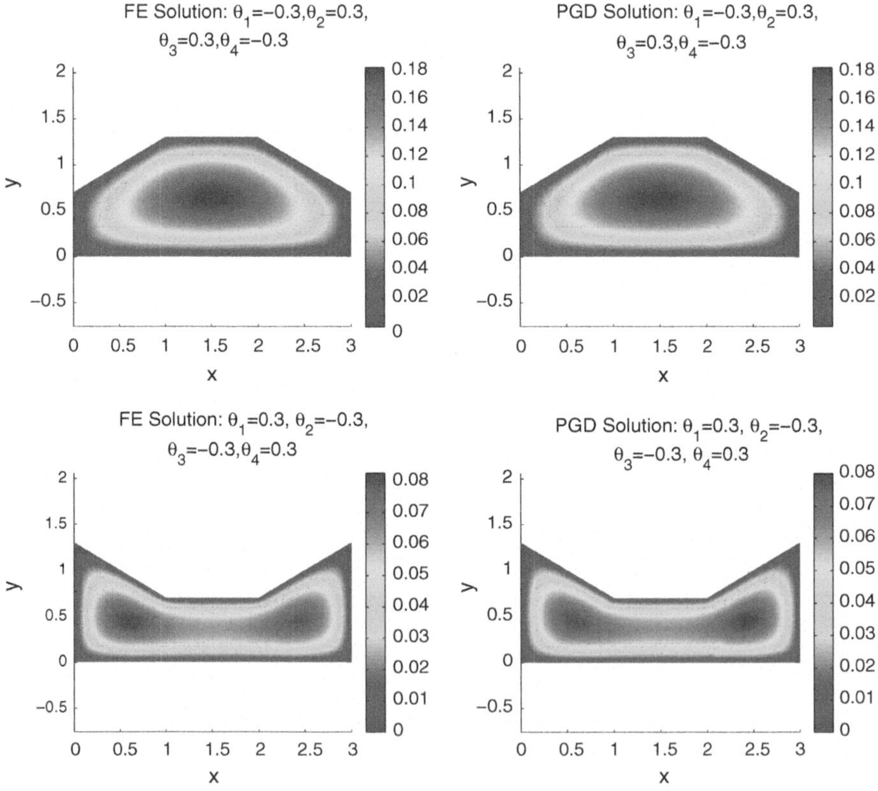

Fig. 8.9 FE (*left*) versus particularized PGD (*right*) solutions in two other geometries

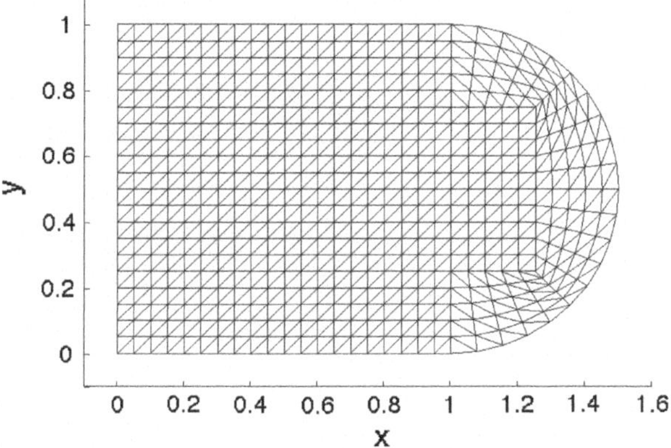

Fig. 8.10 Reference geometry and mesh considered for approximating the spatial functions involved in the separated representation

Thus, the unit square Ω^{r1} results in the rectangle $\Omega^1 =]a, l[\times]0, h[$, with $l = L \cdot \theta_1$ and $h = 2 \cdot r = 2 \cdot R \cdot \theta_2 = H \cdot \theta_2$. The coordinates transformation reads:

$$\begin{cases} x = X \cdot \theta_1 \\ y = Y \cdot \theta_2 \end{cases}, \tag{8.25}$$

that results in the diagonal Jacobian

$$\mathbf{J}_1 = \begin{pmatrix} \theta_1 & 0 \\ 0 & \theta_2 \end{pmatrix}, \tag{8.26}$$

that implies a single product in the expression of $1/\det \mathbf{J}_1$.

On the other hand, when considering the circle, the transformation of $\mathscr{C}(\mathbf{X}_c; R)$ into $\mathscr{C}(\mathbf{x}_c; r)$ ($\mathbf{X}_c$ and $\mathbf{x}_c$ being the initial and final positions of the circle centre respectively) can be considered the combination of a change of the radius, without changing the position of its centre, plus a translation. Because the translation does not appear in the Jacobian, we focus on the circle dilatation.

This transformation reads:

$$\mathbf{x} = \mathbf{X}_c + \theta_2 \cdot (\mathbf{X} - \mathbf{X}_c), \tag{8.27}$$

where the centre is assumed unchanged.

We can notice that this relation implies

$$(\mathbf{x} - \mathbf{X}_c) = \theta_2 \cdot (\mathbf{X} - \mathbf{X}_c) \rightarrow r^2 = \theta^2 \cdot R^2. \tag{8.28}$$

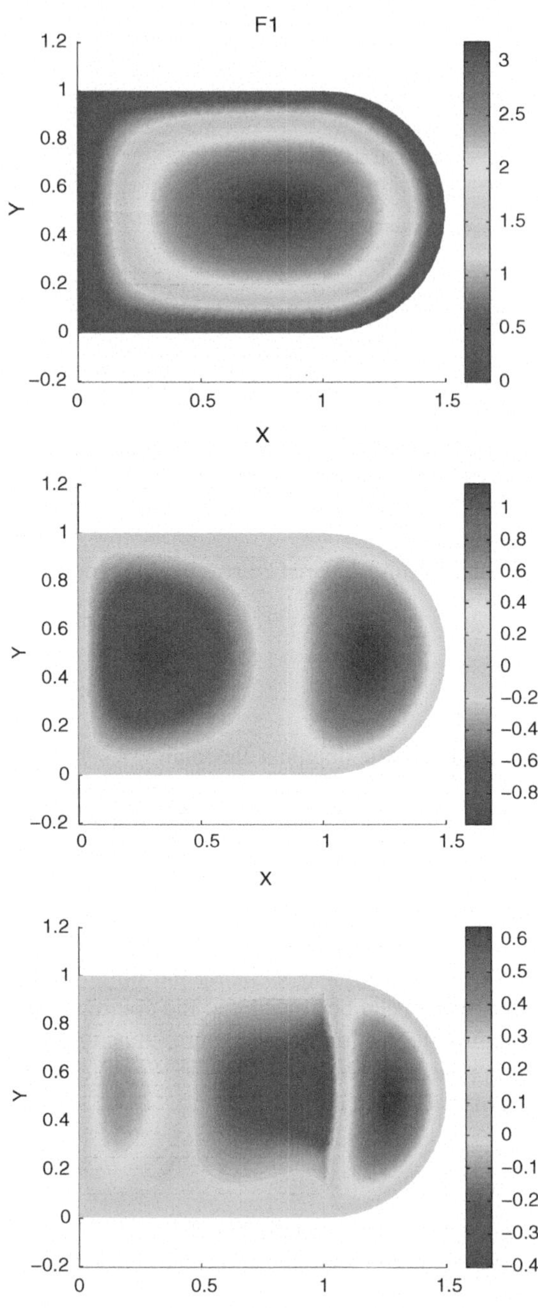

Fig. 8.11 Most significant spatial modes

Fig. 8.12 Most significant modes $G_i(\theta_1, \theta_2)$

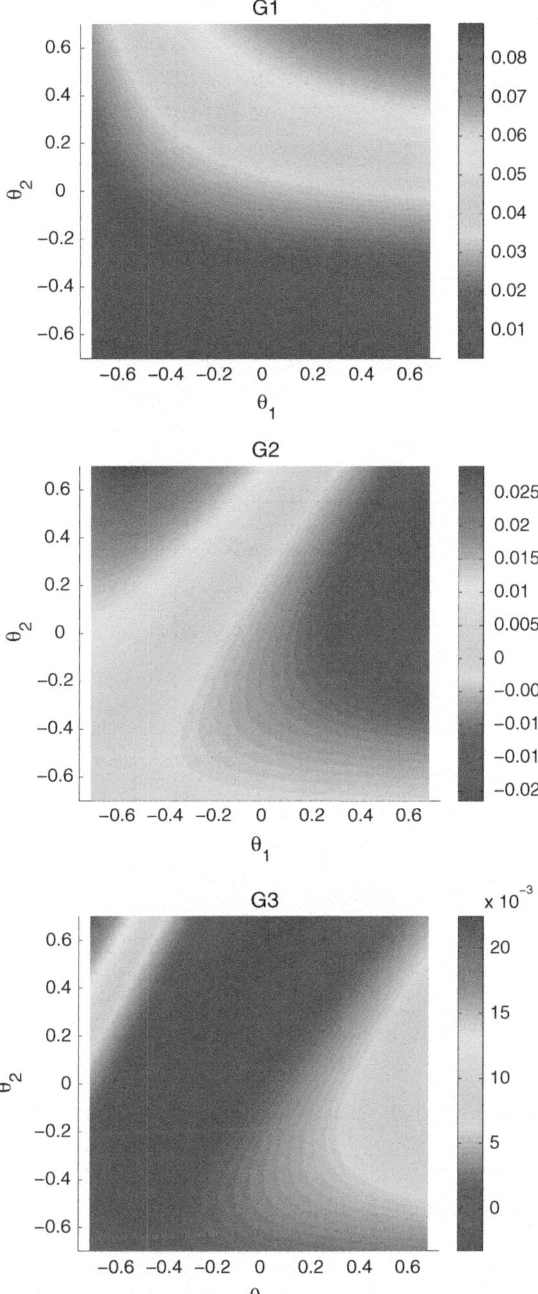

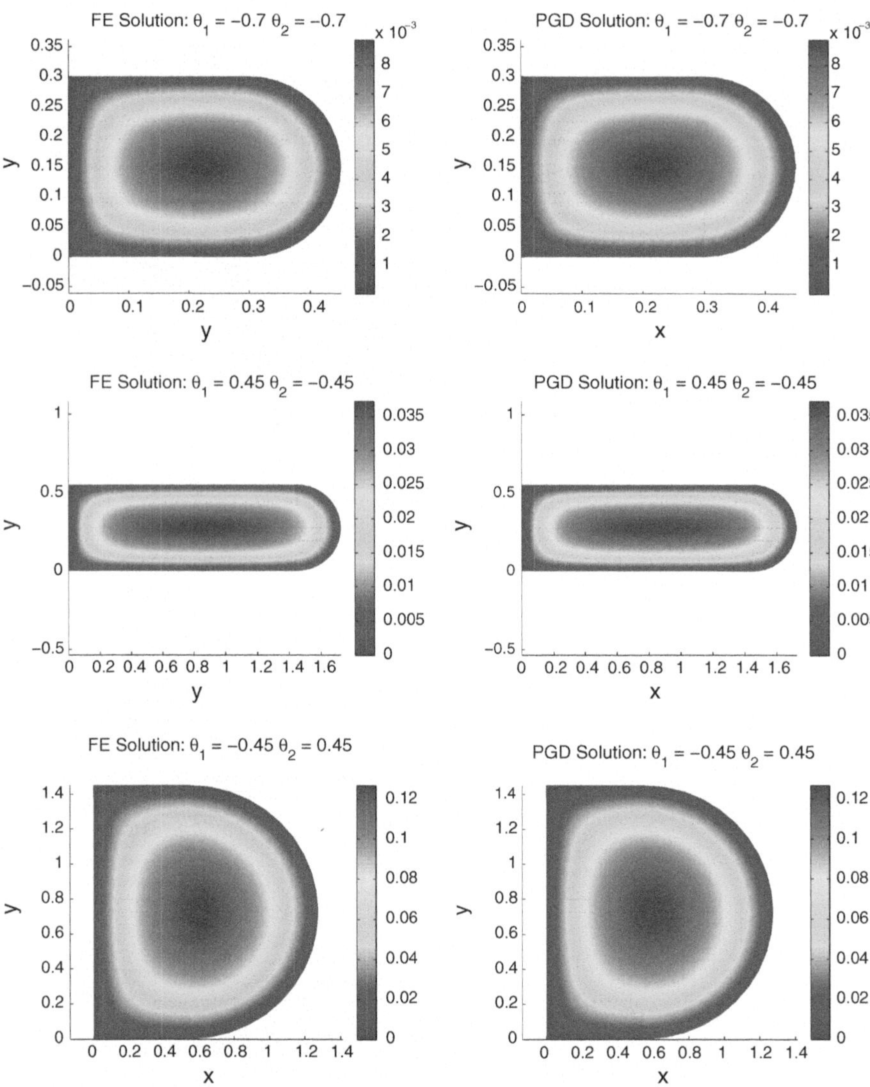

Fig. 8.13 FE (*left*) versus particularized PGD (*right*) solutions in two other geometries

Thus, the circle dilatation transformation results in the diagonal Jacobian

$$\mathbf{J}_2 = \begin{pmatrix} \theta_2 & 0 \\ 0 & \theta_2 \end{pmatrix}. \tag{8.29}$$

The resulting solution separated representation involves 20 terms, even if with only six the error is lower than 1 %:

$$u(\mathbf{x}, \theta_1, \theta_2) \approx \sum_{i=1}^{i=20} F_i(\mathbf{x}) \cdot G_i(\theta_1, \theta_2). \tag{8.30}$$

Functions $G_i(\theta_1, \theta_2)$ were approximated in the parametric domain $[-0.7, 0.7] \times [-0.7, 0.7]$ by using a uniform mesh consisting of 30×30 nodes. On the other hand, functions of space $F_i(\mathbf{x})$ were approximated by using linear finite elements on the mesh shown in Fig. 8.10.

Figure 8.11 depicts the three most significant functions $F_i(\mathbf{x})$ and Fig. 8.12 the three most significant functions $G_i(\theta_1, \theta_2)$.

Finally, Fig. 8.13 depicts the FE solutions and the particularized parametric PGD solutions for other different geometries. Again, both solutions are in perfect agreement.

8.4 Conclusions

This work constitutes a first attempt to consider parametric models in which the parameters related to complex geometries are introduced into the model as new extra-coordinates. The resulting multidimensional model can be efficiently solved by invoking a proper generalized decomposition that allows circumventing the curse of dimensionality.

The difficulties related to the separated representation of the inverse of the Jacobian of the transformation have been alleviated by considering simpler mappings, such as the ones associated with linear triangles that in many cases allow for exact separated representations of the inverse of the Jacobian.

The consideration of geometrical parameters coming for a CAD description or even the ones related to a isogeometric description of the domain boundary constitutes some of the works in progress.

Chapter 9
PGD Based Dynamic Data Driven Application Systems

Dynamic Data-Driven Application Systems—DDDAS—appear as a new paradigm in the field of applied sciences and engineering, and in particular in simulation-based engineering sciences. By DDDAS we mean a set of techniques that allows us to link simulation tools with measurement devices for real-time control of systems and processes. In this work a novel simulation technique is developed with an eye towards its use in the field of DDDAS. The main novelty of this technique relies in the consideration of parameters of the model as new dimensions in the parametric space. Such models often live in highly multidimensional spaces suffering the so-called curse of dimensionality. To avoid this problem related to mesh-based techniques, in this work an approach based upon the Proper Generalized Decomposition (PGD)—is developed, which is able to circumvent the redoubtable curse of dimensionality. The approach thus developed is composed by a marriage of DDDAS concepts and a combination of PGD "off-line" computations, linked to "on-line" post-processing. In this work we explore some possibilities in the context of process control, malfunctioning identification and system reconfiguration in real time, showing the potentialities of the technique in real engineering contexts.

9.1 Introduction: Dynamic Data-Driven Application Systems

Traditionally, Simulation-based Engineering Sciences (SBES) relied on the use of static data inputs to perform the simulations. These data could be parameters of the model(s) or boundary conditions, outputs at different time instants, etc., traditionally obtained through experiments. The word static is intended here to mean that these data could not be modified during the simulation.

A new paradigm in the field of Applied Sciences and Engineering has emerged in the last decade. Dynamic Data-Driven Application Systems (DDDAS) constitute nowadays one of the most challenging applications of SBES. By DDDAS we mean a set of techniques that allow the linkage of simulation tools with measurement devices for real-time control of simulations and applications. As defined by

F. Chinesta and E. Cueto, *PGD-Based Modeling of Materials, Structures and Processes*, ESAFORM Bookseries on Material Forming, DOI: 10.1007/978-3-319-06182-5_9,

the U.S. National Science Foundation, "DDDAS entails the ability to dynamically incorporate additional data into an executing application, and in reverse, the ability of an application to dynamically steer the measurement process" [1].

The term Dynamic Data-Driven Application System was coined by Darema in a NSF workshop on the topic in 2000 [2]. The document that initially put forth this initiative stated that DDDAS constitute "application simulations that can dynamically accept and respond to 'online' field data and measurements and/or control such measurements. This synergistic and symbiotic feedback control loop among applications, simulations, and measurements is a novel technical direction that can open new domains in the capabilities of simulations with a high potential pay-off, and create applications with new and enhanced capabilities. It has the potential to transform the way science and engineering are done, and induces a major beneficial impact in the way many functions in our society are conducted, such as manufacturing, commerce, transportation, hazard prediction/management, and medicine, to name a few" [3].

The importance of DDDAS in the forthcoming decades can be noticed from the NSF Blue Ribbon Panel on SBES report [4], that in 2006 included DDDAS as one of the five core issues or challenges in the field for the next decade (together with multiscale simulation, model validation and verification, handling large data and visualization). This panel concluded that "Dynamic data-driven application systems will rewrite the book on the validation and verification of computer predictions" and that "research is needed to effectively use and integrate data-intensive computing systems, ubiquitous sensors and high-resolution detectors, imaging devices, and other data-gathering storage and distribution devices, and to develop methodologies and theoretical frameworks for their integration into simulation systems" [4]. Moreover, the NSF believes that "...The DDDAS community needs to reach a critical mass both in terms of numbers of investigators, and in terms of the depth, breadth and maturity of constituent technologies..." [1].

A DDDAS includes different constituent blocks:

1. A set of (possibly) heterogeneous simulation models.
2. A system to handle data obtained from both static and dynamic sources.
3. Algorithms to efficiently predict system behaviour by solving the models under the restrictions set by the data.
4. Software infrastructure to integrate the data, model predictions, control algorithms, etc.

Almost a decade after the establishment of the concept, the importance of the challenge is better appreciated. As can be noticed, it deals with very different and transversal disciplines: from simulation techniques, numerical issues, control, modeling, software engineering, data management and telecommunications, among others. The three different blocks of interactions concern: (i) The one between human systems and the simulation, (ii) The simulation interaction with the physical system and (iii) The simulation and the hardware/ data infrastructure. Physical systems operate at very different time scales: from 10^{-20} Hz for cosmological systems to 10^{20} Hz for problems at the atomic scales. Humans, however, can be considered as a system operating at rates from 3 to 500 Hz in haptic devices for instance to transmit realistic touch sensations. A crucial aspect of DDDAS is that of real-time simulation. This means that the simulations must run at the same time (or faster) than data are collected. While this is not always true (as in weather forecasting, for instance, where collected data are usually incorporated to the simulations after long time periods), most applications require different forms of real-time simulations. In haptic surgery simulators, for instance, the simulation result, i.e., forces acting on the surgical tool, must be translated to the peripheral device at a rate of 500 Hz, which is the frequency of the free hand oscillation. In other applications, such as some manufacturing processes, the time scales are much bigger, and therefore real-time simulations can last for seconds or minutes.

As can be noticed from the introduction above, DDDAS can revolutionize the way in which simulation will be done in the next decades. No longer will a single run of a simulation will be considered as a way of validating a design on the basis of a static data set [4].

While research on DDDAS should involve applications, mathematical and statistical algorithms, measurement systems, and computer systems software methods, our work focuses on the development of mathematical and statistical algorithms for the simulation within the framework of such a system. In brief, we will show here how to incorporate PGD techniques into the field, allowing the performance of faster simulations, able to cope with uncertainty, multiscale phenomena, inverse problems and many other features that will be discussed. We revisit the motivation and the key ideas of such a PGD approach in the next sections.

9.1.1 When the Solution of Many Direct Problems is Needed: The Many-Query Context

An important issue encountered in DDDAS, related to process control and optimization, inverse analysis, etc., lies in the necessity of solving many direct problems. Thus, for example, process optimization implies the definition of a cost function and the search of optimum process parameters, which minimize the cost function. In most engineering optimization problems the solution of the model is the most expensive step. Real-time computations with zero-order optimization techniques can not be envisioned except for very particular cases. The computation of sensitivity matrices and adjoint approaches also hampers fast computations. Moreover, global minima are only ensured under severe conditions, which are not (or cannot be) verified in problems of engineering interest. There are many strategies to update the set of design parameters and the interested reader can find most of them in books focusing on optimization procedures. Our interest here is not the discussion on optimization strategies, but pointing out that standard optimization strategies need for numerous direct solutions of the problem that represents the process, one solution for each tentative choice of the process parameters, plus those required for sensitivity.

As already discussed in previous paragraphs, the solution of the model is a tricky task that demands for important computational resources and usually implies extremely large computing times. Usual optimization procedures are inapplicable under real-time constraints because they need for numerous solutions. The same issues are encountered when dealing with inverse analysis in which material or process parameters are expected to be identified from numerical simulation, by looking for the unknown parameters such that the computed fields strongly agree with the ones measured experimentally. However, some previous references exist on the treatment of problems that require extensive solution procedures for different parameter values (the so-called *many-query context*).

9.1.2 Towards Generalized Parametric Modeling

One possibility for solving many problems very fast (many-query problems) consists of using some kind of model order reduction based on the use of reduced basis [5, 6]. In these works authors proved the capabilities of performing real-time simulation even using handheld, deployed, devices, as smartphones for instance. The tricky point in such approaches is the construction of such a reduced basis and the way of adapting them when the system explores regions far from the ones considered in the construction of the reduced model. Solutions to this issue exist and others are being developed to fulfill with real-time requirements.

Multidimensionality offers an alternative getaway to avoid too many direct solutions. It consists of introducing all sources of variability as extra-coordinates into the model and then solving the resulting multidimensional model only once to have access to the whole envelope of solutions. PGD technology allows to circumvent efficiently the curse of dimensionality. In our modest opinion it could represent a new paradigm in computational mechanics. Thus, moving loads in structural mechanics, geometrical parameters in shape optimization, material parameters in material characterization, boundary conditions in inverse analysis or process optimization, etc., can be treated as extra-coordinates to compute *off-line* multidimensional parametric solutions that could then be used *on-line*, running in real time. These general solutions computed off-line can be introduced in very light computing devices, as for example smartphones, opening an unimaginable field of applications that Fig. 9.1 caricatures (Source: http://es.wikipedia.org\discretionary-/wiki/Archivo:UPM-CeSViMa-SupercomputadorMagerit.jpg). This methodology constitutes in our opinion a new paradigm for real-time simulation.

9.2 Proper Generalized Decomposition for a Parametric Model of a Material Flowing in a Heated Die

In this section the main ideas related to casting the model into a multidimensional framework, followed by on-line process optimization, are introduced. For the sake of clarity in what follows we consider the thermal model related to a material flowing

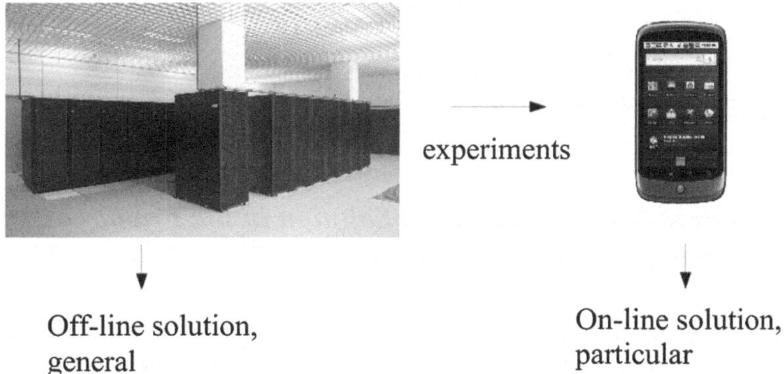

Fig. 9.1 "Off-line" solution of a general enough parametric model and "on-line" particularization of such a general solution in a particular context

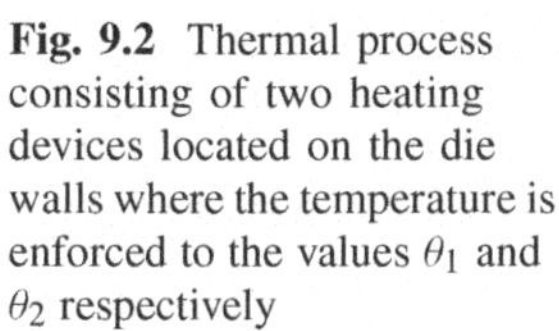

Fig. 9.2 Thermal process consisting of two heating devices located on the die walls where the temperature is enforced to the values θ_1 and θ_2 respectively

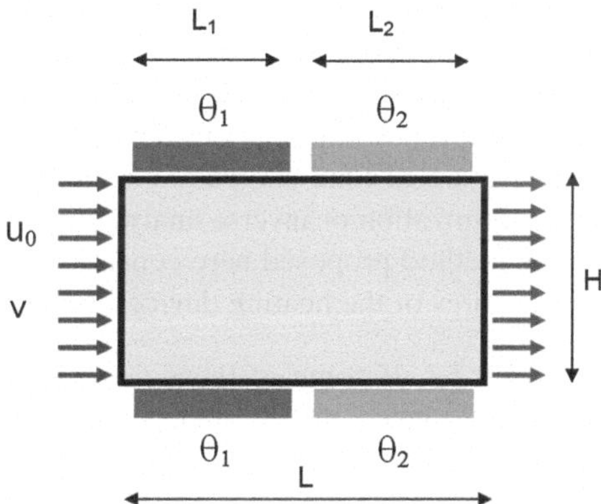

into a heated die (roughly corresponding to a very academic description of pultrusion processes). Despite the apparent simplicity, the strategy here described can be extended to address more complex scenarios.

The assumed 2-dimensional thermal process is sketched in Fig. 9.2. Material flows with a velocity $\mathbf{v}$ within a die Ω of length L and width H. The temperature of the material at the die entrance is u_0. The die is equipped with two heating devices (resistors) of lengths L_1 and L_2 respectively, whose temperatures θ_1 and θ_2 respectively, can be adjusted within an interval $[\theta_{\text{min}}, \theta_{\text{max}}]$.

The steady-state temperature field $u(\mathbf{x})$ at any point of the die $\mathbf{x} = (x, y) \in \Omega \subset \mathbb{R}^2$ can be obtained from the solution of the 2D heat transfer equation that involves in this case both advection and diffusion mechanics as well as a possible source term Q. The velocity field under these assumptions is everywhere unidirectional, i.e., $\mathbf{v}^T = (0, v)$. The steady-state heat transfer equation thus reduces to:

$$\rho c\left(v\frac{\partial u}{\partial x}\right) = k\Delta u + Q, \tag{9.1}$$

where k is the thermal conductivity, ρ is the density and c is the specific heat.

9.2.1 Building Up the Parametric Solution in the Framework of a Multi-dimensional Model

The die is equipped with two heating devices as depicted in Fig. 9.2 whose temperatures constitute the process parameters to be optimized and, eventually, controlled. For the sake of simplicity the internal heat generation Q is assumed constant, as well as the velocity v and the inlet temperature u_0.

Different values of prescribed temperatures at both heating devices can be considered. The resulting 2D heat transfer equation can be then solved. As noted earlier, optimization or inverse identification will require many direct solutions or, as named in the introduction, static data computations. Obviously, when the number of the process parameters involved in the model is increased, standard approaches fail to compute optimal solutions in a reasonable time. Thus, for a large number of process parameters, real-time computations are precluded and, moreover, performing "on-line" optimization or inverse analysis is a challenging issue.

The method proposed here consists in introducing both process parameters, i.e., temperatures of the heating devices, θ_1 and θ_2, as extra-coordinates.

Remark 9.2.1 If some of these desired extra-coordinates are fields depending on other coordinates instead independent parameters (temperatures of the heating devices evolving in time, source term evolving in time and/or space...) prior to introduce them as extra-coordinates one should parametrize such evolutions in an appropriate manner, and finally introduce the coefficients involved in those parametrizations as extra-coordinates as considered in [7].

Other parameters such as $\rho, c, v, k, \ldots$, can be set as extra-coordinates as well. The temperature field can thus be computed at each point and for any possible value of the temperatures θ_1 and θ_2 (within the already fixed intervals already mentioned). As soon as this multidimensional solution $u(x, y, \theta_1, \theta_2)$ is available, it is possible to particularize it for any value of the process parameters without the need for further executions of the code. Thus, optimization procedures can proceed with only knowledge of an "off-line" pre-computed parametric solution.

To circumvent the curse of dimensionality related to the high-dimensional space in which the temperature field $u(x, y, \theta_1, \theta_2)$ is defined—which we assume to be four-dimensional for the ease of exposition—we consider a separated representation of that field:

$$u(x, y, \theta_1, \theta_2) \approx \sum_{i=1}^{N} F_i(x, y)\ \Theta_i^1(\theta_1)\ \Theta_i^2(\theta_2) \tag{9.2}$$

where all the functions involved in such separated representation can be computed by applying the Proper Generalized Decomposition technique, described previously.

Remark 9.2.2 Because of the geometrical simplicity of Ω that can be written as $\Omega = \Omega_x \times \Omega_y$, we could consider a fully separated representation of the unknown field that now reads:

$$u(x, y, \theta_1, \theta_2) \approx \sum_{i=1}^{N} X_i(x) \cdot Y_i(y) \cdot \Theta_i^1(\theta_1) \cdot \Theta_i^2(\theta_2) \tag{9.3}$$

this fully separated representation can be also applied in complex domains as proved in [8], however when the physical domain is complex the most natural representation is that given by Eq. (9.2).

The prescribed essential boundary conditions are written:

$$\begin{cases} u(x = 0, y, \theta_1, \theta_2) = u_0, \\ u(x \in I_1, y = 0 \text{ or } y = H, \theta_1, \theta_2) = \theta_1, \\ u(x \in I_2, y = 0 \text{ or } y = H, \theta_1, \theta_2) = \theta_2, \\ \frac{\partial u}{\partial x}(x = L, y, \theta_1, \theta_2) = 0, \end{cases} \tag{9.4}$$

where I_1 and I_2 are the intervals of the x coordinate where the heating devices of length L_1 and L_2 respectively are defined. A null heat flux is assumed in the remaining part of the domain boundary. Thus, the temperature field u depends on four different coordinates, the two space coordinates (x, y) and the two temperatures prescribed in both regions on the die wall. Parameters θ_1 and θ_2 now take values in the intervals $\mathscr{I}_1$ and $\mathscr{I}_2$ respectively.

The case of essential (Dirichlet) boundary conditions as parameters of the model deserves some comments. Non-homogeneous essential boundary conditions in PGD methods are usually treated by means of a simple change of variable;

$$u = \psi + z, \tag{9.5}$$

where ψ is a function verifying essential boundary conditions. This leads to a problem in the z variable with homogeneous boundary conditions. Efficient construction of ψ functions in the framework of PGD approximations has been deeply analyzed in [8]. We refer the interested reader to this reference for further details. This function ψ can, for the problems addressed here, be expressed in separated form

$$\psi(x, y, \theta_1, \theta_2) = \sum_{i=1}^{3} F_i(x, y)\ \Theta_i^1(\theta_1)\ \Theta_i^2(\theta_2), \tag{9.6}$$

where each functional product is used to impose initial conditions and the two non-homogeneous essential boundary conditions, one for each heater position. These functions are depicted in Fig. 9.3.

The resulting PGD approximation reads

$$u(x, y, \theta_1, \theta_2) \approx \psi(x, y, \theta_1, \theta_2) + \sum_{i=4}^{N} F_i(x, y)\ \Theta_i^1(\theta_1)\ \Theta_i^2(\theta_2) \tag{9.7}$$

Functions F_i, Θ_i^1 and Θ_i^2, $i = 4, \ldots, N$, are determined by solving a sequence of non-linear problems with homogeneous boundary conditions, as described below. Assuming that the first n functional products of Eq. (9.7) have already been computed, we look for the $n + 1$ term:

$$u^{n+1}(x, y, \theta_1, \theta_2) = \sum_{i=1}^{n} F_i(x, y)\ \Theta_i^1(\theta_1)\ \Theta_i^2(\theta_2) + R(x, y)\ S(\theta_1)\ T(\theta_2) \tag{9.8}$$

or, equivalently,

$$u^{n+1}(x, y, \theta_1, \theta_2) = u^n(x, y, \theta_1, \theta_2) + R(x, y)\ S(\theta_1)\ T(\theta_2). \tag{9.9}$$

Note that in the weak form associated to (9.1) the unknown functions R, S, and T are determined and the test functions are

$$\begin{aligned} u^*(x, y, \theta_1, \theta_2) = {} & R^*(x, y)\ S(\theta_1)\ T(\theta_2) + R(x, y)\ S^*(\theta_1)\ T(\theta_2) \\ & + R(x, y)\ S(\theta_1)\ T^*(\theta_2). \end{aligned} \tag{9.10}$$

This approach allows us to determine the unknown functions $R(x, y)$, $S(\theta_1)$ and $T(\theta_2)$ in an alternating directions fixed-point algorithm. In fact, we proceed by determining sequentially each one of these functions, as described below, until reaching convergence. For more details we refer the interested reader to [9].

Remark 9.2.3 In the present case a linear model is assumed. Many problems encountered in science and engineering are however nonlinear. For instance, the heat equation here addressed becomes nonlinear as soon as the dependence of thermal parameters on the temperature field is taken into account.

In the non-linear case additional linearisation techniques are required. Application of standard linearisation strategies within the PGD framework were addressed in [10] and [11], for instance. An even better way to address strongly non-linear models consists in using the LATIN technique [12]. However its extension to the case of multidimensional parametric modeling is not straightforward.

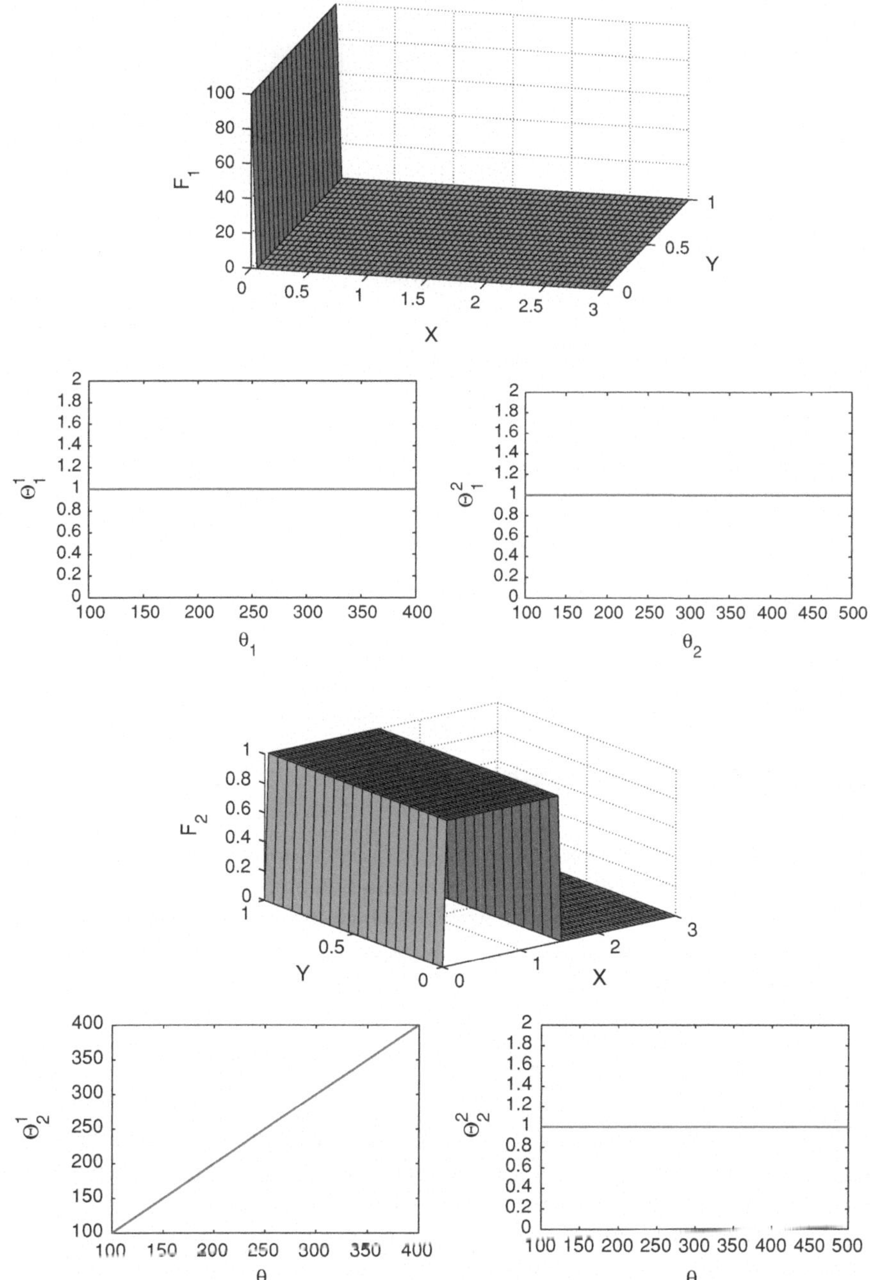

Fig. 9.3 Functions $F_i(x, y)$, $\Theta_i^1(\theta_1)$ and $\Theta_i^2(\theta_2)$ used to approximate initial and boundary conditions

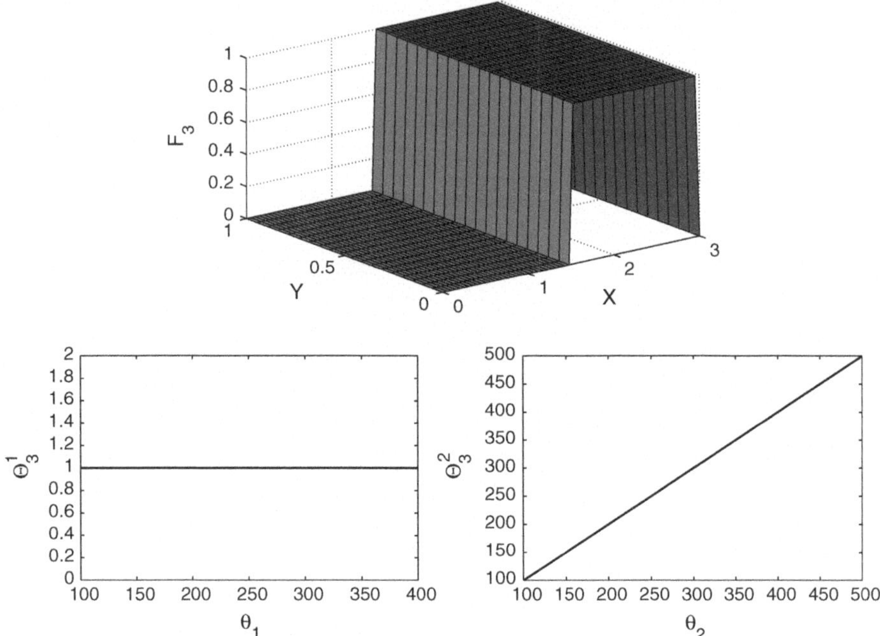

Fig. 9.3 (Continued)

9.2.2 "Off-Line" Optimization Procedure

Optimization procedures look for optimal parameters minimizing an appropriate single- or multi-objective cost function (sometimes subjected to many constraints). In this work we consider a simple scenario, in which the cost function only involves the coldest thermal history of an imaginary material particle traversing the die. It is therefore expressed as:

$$\mathscr{C}(\theta_1, \theta_2) = \frac{1}{2}\left(\int_0^L u\left(x, \tfrac{H}{2}, \theta_1, \theta_2\right) dx - \beta\right)^2, \tag{9.11}$$

where β denotes the optimal value of the thermal history able to ensure a certain material transformation (curing). Values lower than β imply that the material has not received the necessary amount of heat, whereas values higher than β imply an unnecessary extra-heating.

Obviously, optimal process parameters θ_1^{opt} and θ_2^{opt} must be calculated by minimizing the cost function. There exist many techniques for such minimization. The interested reader can refer to any book on optimization. Many of them proceed by evaluating the gradient of the cost function and then moving on that direction. The gradient computation involves the need for computing first derivatives of the cost function with respect to the process parameters. Other techniques involve the calculation of second derivatives. To this end, one should calculate the derivatives of the problem solution with respect to the optimization parameters.

It is important to note that separated representations of the process parameters drastically simplifies this task because, since the solution depends explicitly on the parameters, its derivation is straightforward, namely,

$$\frac{\partial u}{\partial \theta_1}(x, y, \theta_1, \theta_2) \approx \sum_{i=1}^{N} F_i(x, y) \frac{\partial \Theta_i^1}{\partial \theta_1}(\theta_1)\, \Theta_i^2(\theta_2),$$

and

$$\frac{\partial u}{\partial \theta_2}(x, y, \theta_1, \theta_2) \approx \sum_{i=1}^{N} F_i(x, y)\, \Theta_i^1(\theta_1) \frac{\partial \Theta_i^2}{\partial \theta_2}(\theta_2).$$

Note that second derivatives can be also similarly obtained. The calculation of the solution derivatives is a tricky point when proceeding from standard discretization techniques because the parametric dependency of the solution is, in general, not explicit.

Moreover, the separated, rank-N, representation of the solution see (9.2), further simplifies the expression of the cost function. That is, substituting (9.2) in (9.11), induces

$$\mathscr{C}(\theta_1, \theta_2) = \frac{1}{2}\Big(\sum_{i=1}^{N} \alpha_i\, \Theta_i^1(\theta_1)\, \Theta_i^2(\theta_2) - \beta\Big)^2 \tag{9.12}$$

where $\alpha_i = \int_0^L F_i(x, \frac{H}{2})dx$, and the different derivatives of the cost function becomes:

$$\begin{cases} \dfrac{\partial \mathscr{C}}{\partial \theta_1}(\theta_1, \theta_2) = \Big(\sum_{i=1}^{N} \alpha_i\, \Theta_i^1(\theta_1)\, \Theta_i^2(\theta_2) - \beta\Big)\Big(\sum_{i=1}^{N} \alpha_i\, \frac{\partial \Theta_i^1}{\partial \theta_1}(\theta_1)\, \Theta_i^2(\theta_2)\Big), \\ \dfrac{\partial \mathscr{C}}{\partial \theta_2}(\theta_1, \theta_2) = \Big(\sum_{i=1}^{N} \alpha_i\, \Theta_i^1(\theta_1)\, \Theta_i^2(\theta_2) - \beta\Big)\Big(\sum_{i=1}^{N} \alpha_i\, \Theta_i^1(\theta_1)\, \frac{\partial \Theta_i^2}{\partial \theta_2}(\theta_2)\Big). \end{cases}$$

In the simulations carried out and reported in subsequent sections, the minimization of the cost function was performed by using a Levenberg-Marquardt algorithm, see [13] for further details.

To briefly describe the Levenberg-Marquardt algorithm we introduce the following notation: $\mathbf{p} = (\theta_1, \theta_2)^T$ is the vector that contains the parameters involved in the optimization. We denote by

$$f(\mathbf{p}) = \sum_{i=1}^{N} \alpha_i\, \Theta_i^1(\theta_1)\, \Theta_i^2(\theta_2),$$

such that the cost function (9.12) can be rewritten as:

$$\mathscr{C}(\theta_1, \theta_2) = \frac{1}{2}\big(f(\mathbf{p}) - \beta\big)^2$$

The algorithm is initialized with some set of parameters $\mathbf{p}^0 = (\theta_1^0, \theta_2^0)^T$. It then proceeds by repeating the following steps until convergence:

1. Compute the Jacobian at iteration k, $\mathbf{J}^k$, namely:

$$\mathbf{J}^k = \frac{\partial f}{\partial \mathbf{p}^T}(\theta_1^k, \theta_2^k)$$

2. Compute the residual at iteration k, r^k:

$$r^k = f(\mathbf{p}^k) - \beta. \tag{9.13}$$

3. Update the state vector $\mathbf{p}^{k+1} = (\theta_1^{k+1}, \theta_2^{k+1})$ according to:

$$\mathbf{p}^{k+1} = \mathbf{p}^k - \left(\left(\mathbf{J}^k\right)^T \mathbf{J}^k + \lambda \mathbf{I}\right)^{-1} \cdot \left(\mathbf{J}^k\right)^T \cdot r^k, \tag{9.14}$$

   where $\mathbf{I}$ is in the present case the 2×2 unit matrix.
4. If the cost function decreases very fast we can reduce the relaxation coefficient λ and then the Levenberg-Marquardt algorithm approaches the Newton one. On the other hand, if the cost function decreases slowly we can increase the value of λ to approach a classical gradient strategy. We consider a standard adaptation of such a coefficient.

In the model considered in this work the expressions of the Jacobian and the residual can be computed explicitly.

9.2.3 Defining the Process Control

Once the optimal parameters θ_1^{opt} and θ_2^{opt} are determined, the general solution, see (9.2), can be evaluated for those optimal values of the process parameters, namely

$$u(x, y, \theta_1^{\text{opt}}, \theta_2^{\text{opt}}) \approx \sum_{i=1}^{N} F_i(x, y)\, \Theta_i^1(\theta_1^{\text{opt}})\, \Theta_i^2(\theta_2^{\text{opt}})$$

to obtain the temperature field everywhere in the domain Ω.

For the sake of clarity, in what follows, we illustrate the dynamic data driven thermal model trough an academic idealization. We consider the thermal model described above, see Fig. 9.2, with the following values of the different parameters (all units defined in the metric system): $\rho = 1$, $c = 1$, $k = 1$, $Q = 50$, $v = 1$, $u_0 = 100$, $L = 3$, and $H = 1$. The first resistor acts in the interval $I_1 = [0.2, 1.4]$ whereas the second one is defined by $I_2 = [1.6, 2.8]$, both having the same length $L_1 = L_2 = 1.2$.

To solve the parametric model one needs to approximate the functions $F_i(\mathbf{x})$ involved in the separated representation of the solution (9.2), as well as functions $\Theta_i^1(\theta_1)$ and $\Theta_i^2(\theta_2)$. Space functions are approximated by using a finite element mesh composed of four-node quadrilateral elements, on a uniform nodal distribution composed of 60×20 nodes in the x and y direction respectively. Functions depending on the process parameters θ_1 and θ_2 are approximated by two uniform 1D linear finite meshes (300 nodes each) over the interval of variation of these parameters ($\mathscr{I}_1 = [100, 400]$ and $\mathscr{I}_2 = [100, 500]$).

The parametric solution $u(\mathbf{x}, \theta_1, \theta_2)$ is computed by using the Proper Generalized Decomposition strategy illustrated in Sect. 9.2.1. This solution implies 42 functional products in the sum, that is $N = 42$ in Eq. (9.2). From this general solution we compute the optimal process parameters θ_1^{opt} and θ_2^{opt} with respect to the cost function introduced in (9.11) where $\beta = 897$. The convergence of the Levenberg-Marquardt algorithm is reached in only 4 iterations, the optimal values being $\theta_1^{\text{opt}} = 275.8$ and $\theta_2^{\text{opt}} = 353.4$. Figure 9.4 depicts the resulting temperature field related to the optimal process parameters.

Remark 9.2.4 Up to now, the convergence criterion that we use involves norms of the residuals. We have considered in our former works more sophisticated error estimators, as the one based on quantities of interest [14], for instance. Error estimators based on quantities of interest are preferred because they are more adapted to the outputs considered in control strategies. However, at present we have not extended these estimators to the multidimensional parametric modeling. Thus, prior to consider any parametric solution computed off-line in on-line procedures we should proceed to verification and validation.

In order to control the process we could imagine two thermocouples located, for instance, at two points on the axis of symmetry, namely $\mathbf{P}_1 = (1, 0.5)$ and $\mathbf{P}_2 = (2, 0.5)$, see also Fig. 9.5. Obviously, for assumed vanishing validation and verification errors, the "on-line" measurements $\tilde{u}_1$ and $\tilde{u}_2$ will give values (almost) coincident with the predictions of our model. These predictions are easily computed from the separate representation of the solution, i.e.,

$$u^{\text{opt}}(\mathbf{P}_1) = u(1, 0.5, \theta_1^{\text{opt}}, \theta_2^{\text{opt}}) = 286.6$$
$$u^{\text{opt}}(\mathbf{P}_2) = u(2, 0.5, \theta_1^{\text{opt}}, \theta_2^{\text{opt}}) = 352.$$

Under these assumptions the process can be considered working in optimal conditions, that related to the optimal temperatures of both heating devices.

However, if the resistors do not work optimally the thermocouple measurements $\tilde{u}_1$ and $\tilde{u}_2$ will differ from the predicted optimal conditions $u^{\text{opt}}(\mathbf{P}_1)$ and $u^{\text{opt}}(\mathbf{P}_2)$.

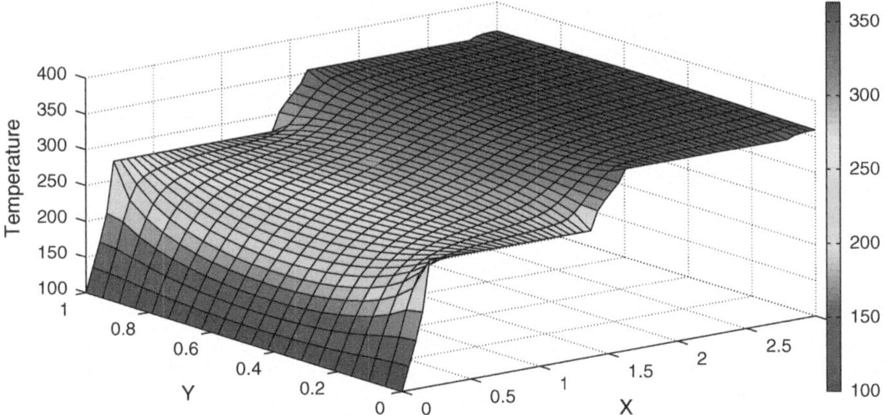

Fig. 9.4 Multi-dimensional solution particularized for the optimal temperature of both heating devices: $u(x, y, \theta_1^{\text{opt}} = 275.8, \theta_2^{\text{opt}} = 353.4)$

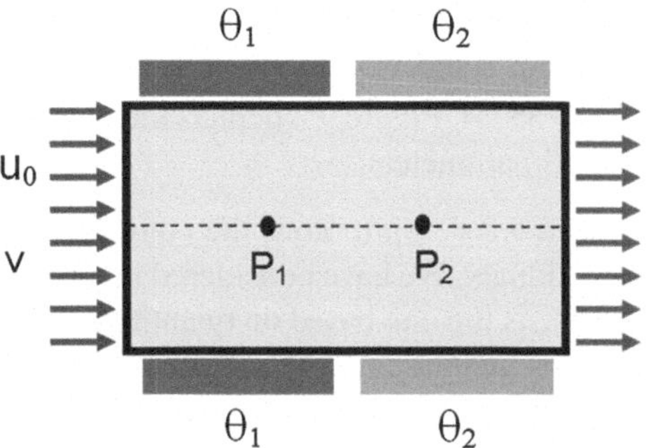

Fig. 9.5 Location of the thermocouples $\mathbf{P}_1$ and $\mathbf{P}_2$ whose measurements serve for controlling the process, identify malfunctioning devices and reconfigure the system after a breakdown

Under these circumstances we could infer a breakdown in the system. The most important point is, however, the identification of the malfunction and the reconfiguration of the system taking into account the real state of the system. These questions are addressed in the next section.

9.3 Simulating a Breakdown Scenario

Consider now a breakdown scenario. Solving the breakdown means: (i) Identifying the broken device and (ii) Constructing an appropriate strategy for process reconfiguration. Both tasks should be performed as fast as possible and using the lightest computational resources (e.g., a smartphone) in order to be able to do real-time decision making independently of the expertise of the personnel in charge of the process.

9.3.1 "On-Line" Inverse Analysis

For illustration purposes, we assume that the second resistor only achieves a fraction of the desired temperature. That is, the device is only able to prescribe a temperature of $\theta_2^{\text{brk}} = 0.4\theta_2^{\text{opt}} = 141.4$, instead of the optimal one (the slight gap is due to the fact of considering few iterations in order to fulfill real time requirements). Under these circumstances, both thermocouples will indicate temperatures at $\mathbf{P}_1$ and $\mathbf{P}_2$ equal to

$$\begin{aligned} u^{\text{brk}}(\mathbf{P}_1) &= u(1, 0.5, \theta_1^{\text{opt}}, \theta_2^{\text{brk}}) = 254 \\ u^{\text{brk}}(\mathbf{P}_2) &= u(2, 0.5, \theta_1^{\text{opt}}, \theta_2^{\text{brk}}) = 165. \end{aligned} \tag{9.15}$$

These values are obtained from the representation of the general solution, see (9.2), since validation and verification errors are assumed negligible.

To reproduce the practical scenario, it is assumed that a decision must be taken with only the information of the thermocouple temperatures, namely (9.15). That is, it is not known which device is malfunctioning and, of course, that the precise value of the actually prescribed temperature is θ_2^{brk}. In fact, the only available datum is (9.15) which clearly indicates that $u^{\text{opt}}(\mathbf{P}_1) \neq u^{\text{brk}}(\mathbf{P}_1)$ and $u^{\text{opt}}(\mathbf{P}_2) \neq u^{\text{brk}}(\mathbf{P}_2)$. Under these circumstances, the system should be reconfigured to ensure that the process continues working.

The first step is to determine which are the actual working temperatures. Thus, an inverse analysis is required. The following step, see next subsection, will be to reconfigure the process in order to fulfill the cost function, i.e., to impose the desired thermal history, see (9.11). Obviously, both steps require a swift resolution if real-time decisions are required.

The "on-line" inverse analysis minimizes the following least-squares problem,

$$\tilde{\mathscr{C}}(\theta_1, \theta_2) = \frac{1}{2}\sum_{i=1}^{2}\left(u^{\text{brk}}(\mathbf{P}_i) - u(x_i, 0.5, \theta_1, \theta_2)\right)^2 \tag{9.16}$$

where x_i for $i = 1, 2$ are the coordinates of the points at which the thermocouples are placed. The Levenberg-Marquardt algorithm, see [13], reaches convergence after three iterations and the estimated temperatures for both heating devices are $\theta_1^{\text{est}} = 261$ and $\theta_2^{\text{est}} = 146$, agree with the considered scenario ($\theta_1^{\text{opt}} = 275.8$, $\theta_2^{\text{brk}} = 141.4$). The inverse identification runs very fast and it only involves cheap calculations, so it could be performed "on-line" on a very light computing devices, such as a smartphone.

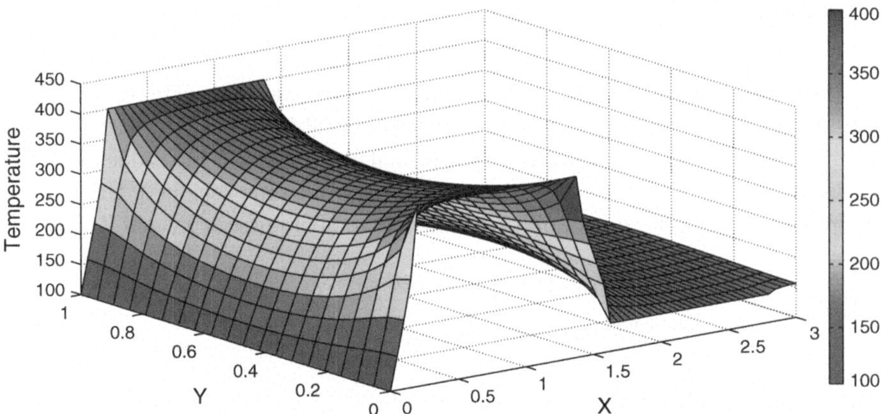

Fig. 9.6 Multi-dimensional solution particularized for the optimal temperature after reconfiguring the system: $u(x, y, \theta_1^*, \theta_2^{\mathrm{est}})$

9.3.2 "On-Line" Process Reconfiguration

Finally, the process is reconfigured to impose the desired thermal history, see (9.11). Obviously, there are many possibilities and strategies. Here, since the second resistor is the one not giving the desired optimal heating, the action consists in keeping the second heating device at its present state (if it is broken this may be the only option), i.e., $\theta_2^{\mathrm{est}} = 146$ and looking for the optimal value of θ_1^* (that seems to be working properly, and therefore admits a reconfiguration), by minimizing the cost function (9.11) for a fixed and known $\theta_2 = \theta_2^{\mathrm{est}} = 146$. This results in

$$\mathscr{C}(\theta_1) = \frac{1}{2}\left(\int_0^L u\left(x, \tfrac{H}{2}, \theta_1, \theta_2^{\mathrm{est}}\right) dx - \beta\right)^2.$$

In three iterations of the Levenberg-Marquardt algorithm convergence is attained to the value $\theta_1^* = 400$. Figure 9.6 depicts the resulting temperature field related to the new optimized process parameters $\theta_1^* = 400$ and $\theta_2^{\mathrm{est}} = 146$.

9.4 Parametric Solution Post-processing by Using Light Computing Platforms

As soon as the parametric solution $u(x, y, \theta_1, \theta_2)$ has been computed only once and off-line, the subsequent processes, i.e., optimization, control and system reconfiguration, only involve very light calculations that could be performed on-line and using handheld computing platforms, like smartphones or tablets.

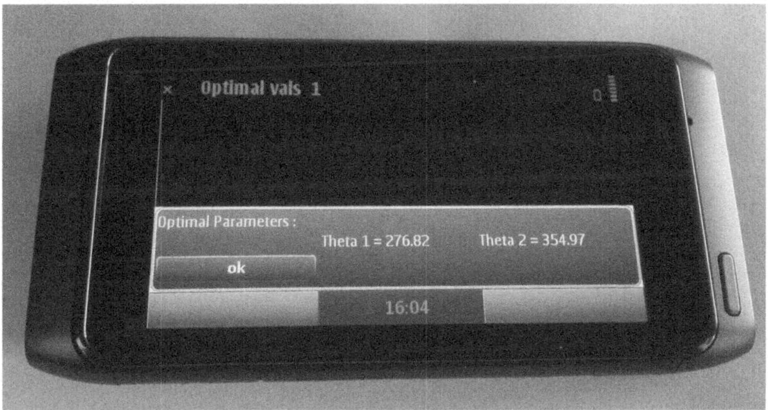

Fig. 9.7 Optimal temperatures of both heating devices

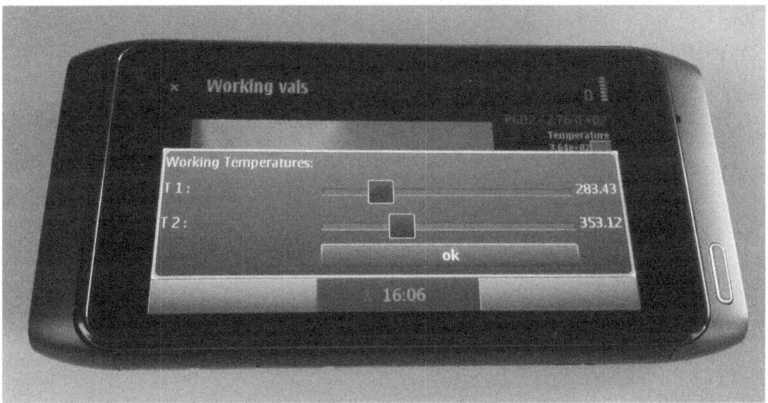

Fig. 9.8 Optimal temperatures at positions $\mathbf{P}_1$ and $\mathbf{P}_2$ when both heaters work at the optimal conditions

To illustrate this capability, we assume that the parametric solution previously considered has already been computed and that it is available in a separated form. This solution can be easily introduced in a smartphone that will perform all the on-line calculations described in the previous sections. Note that modes are stored very efficiently as vectors. Handheld devices should only perform vector multiplications to obtain the solution. Only the most significative modes of the separated representation are retained in order to speed-up the computations. In our applications we considered a Nokia platform, with 256 MB of RAM, 16 GB of internal memory, a 680 MHz ARM 11 CPU and with Symbian3 as operating system.

Figures 9.7–9.13 illustrate all the scenarios analyzed in the previous sections. The smartphone is assumed to be receiving regular updates on the lectures of the thermocouples. Figure 9.7 shows the output for the optimal values of both heating devices $\theta_1^{\text{opt}} = 276.82$ and $\theta_2^{\text{opt}} = 354.97$. They are slightly different from the

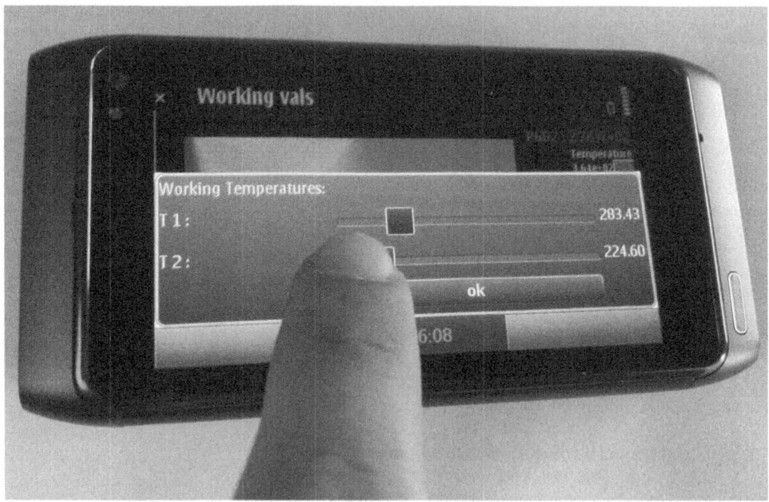

Fig. 9.9 Simulating a failure by considering a temperature at position $\mathbf{P}_2$, different from the optimal one

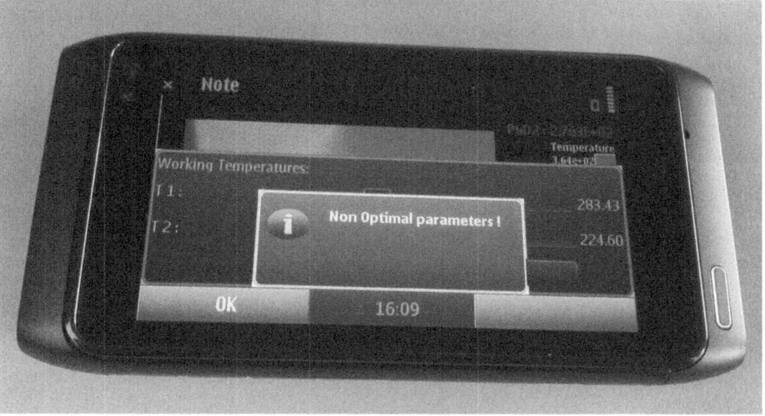

Fig. 9.10 Malfunction alert

ones previously computed because as just argued the parametric solution introduced into the smartphone has been restricted to the most significant modes of the off-line parametric solution to alleviate the on-line calculations performed by the smartphone.

Temperatures at thermocouple positions $\mathbf{P}_1 = (1, 0.5)$ and $\mathbf{P}_2 = (2, 0.5)$ are then computed for optimal temperatures of both resistors. Figure 9.8 shows both temperatures $u^{\text{opt}}(\mathbf{P}_1) = 283.43$ and $u^{\text{opt}}(\mathbf{P}_2) = 353.12$.

A failure scenario is then simulated by considering that the temperatures at positions $\mathbf{P}_1$ and $\mathbf{P}_2$ are not the optimal ones. In fact we consider the scenario defined by $u^{\text{brk}}(\mathbf{P}_1) = 283.43$ and $u^{\text{brk}}(\mathbf{P}_2) = 224.60$ illustrated in Fig. 9.9. The system

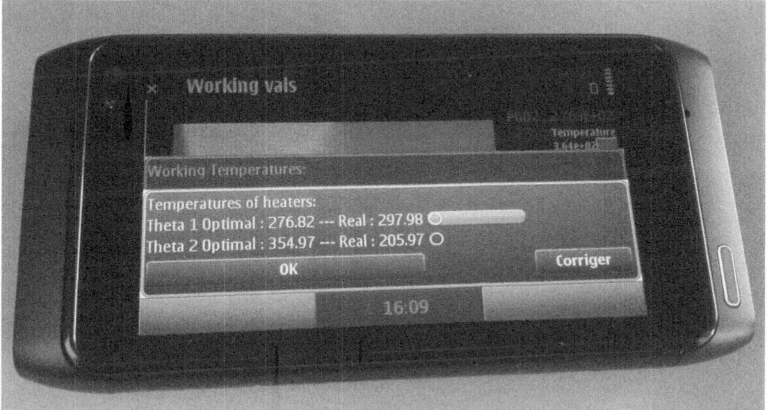

Fig. 9.11 Identification of the real process parameters, i.e. the real temperatures of both resistors

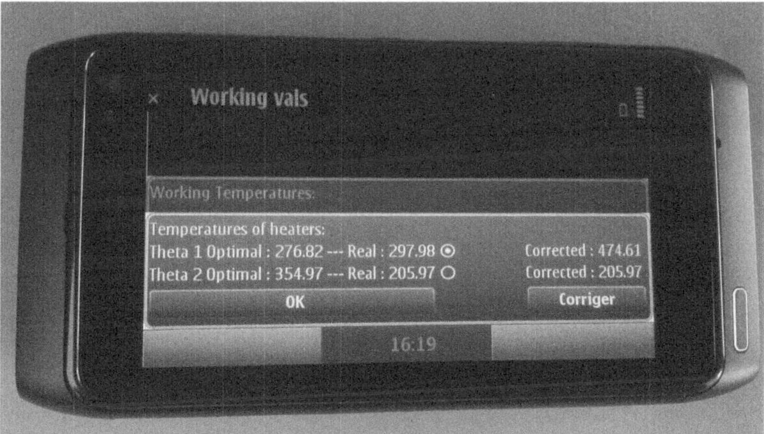

Fig. 9.12 System reconfiguration by adjusting the temperature of the first resistor

identifies the malfunction and displays an alert message as shown in Fig. 9.10. The system then identifies the real temperatures of both heating devices, by applying the inverse strategy previously discussed. Actual temperatures of both resistors are identified as $\theta_1^{est} = 297.98$ and $\theta_2^{est} = 205.97$ instead of the optimal ones $\theta_1^{opt} = 276.82$ and $\theta_2^{opt} = 354.97$ as depicted in Fig. 9.11.

We can decide to reconfigure the system by choosing to change the temperature of one (or eventually both) heating devices. In our case we decide to calculate the new temperature of the first resistor able to ensure optimal process conditions. To this end, we select the first heater as shown in Fig. 9.11. The system then recomputes on-line the optimal temperature of the first heating device: $\theta_1^* = 474.61$ as shown in Fig. 9.12. The new temperature field corresponding to the new operational conditions $u(x, y, \theta_1^* = 474.61, \theta_2^{est} = 205.97)$ is illustrated in Fig. 9.13.

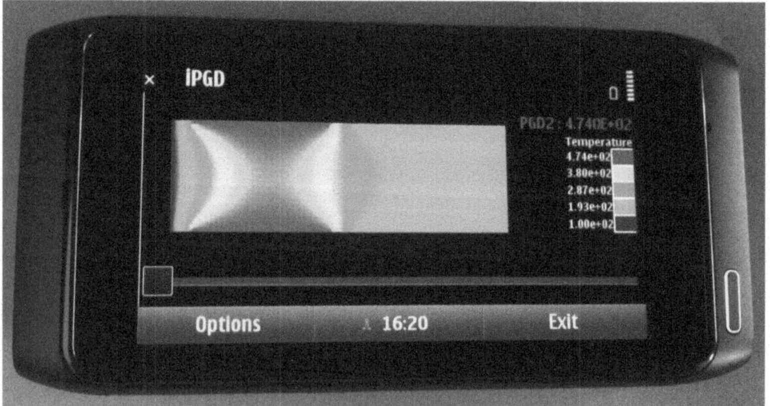

Fig. 9.13 New optimal temperature field related to the reconfigured heating system

9.5 Conclusions

This chapter constitutes a first approach to an analysis of DDDAS based on the use of PGD techniques. The proposed procedure combines first a heavy "off-line" phase. It involves CPU-intensive computations to solve the parametric partial differential equations-based model associated to the industrial process. By introducing all the sources of variability or process parameters as extra-coordinates, the resulting model becomes highly multidimensional. This makes the use of the well experienced mesh-based discretization techniques impossible, due to the curse of dimensionality. However, the use of separated representations within the proper generalized decomposition framework allows us to circumvent this burden in an efficient way.

Once this parametric solution is available, the proposed PGD method proceeds by optimizing the process, still "off-line", by calculating the optimal process parameters in order to minimize an appropriate cost function. It has been emphasized how this optimization procedure benefits from the separated structure of the PGD solution. It helps not only to efficiently store the solution in the form of a series of vectors, but also allows for an extremely fast application of optimization procedures based upon this separated structure, as seen before.

The system response in the event of a breakdown should be computed, however, "on-line" and as fast as possible. Separated representation of the solution built-up by applying the PGD-based solver allows computing explicitly the derivatives of the solution with respect to the process parameters, making the fast calculation of minimization strategies possible. Thus, malfunctioning devices can be identified "on-line" and the systems reconfigured by making some light calculations that we could perform using for example a simple smartphone.

In our opinion the possibility of computing parametric solutions that are then used by simple post-processing opens an unlimited number of potential applications. For instance, uncertainty in the measurements (that has not been considered here), that constitutes an essential ingredient in DDDAS, could eventually be efficiently treated in the PGD framework by considering uncertain parameters as additional dimensions. Worst-case scenarios could be computed in a similar way than the here proposed optimization procedures.

In the numerical, academic, examples here addressed "off-line" calculations needed two minutes of computing time (using Matlab©, a standard laptop and an in-house simulation code) whereas all the "on-line" calculations were performed in 0.0015 s. The examples here addressed are too simple to be conclusive, but at least, they prove the pertinence of the proposed approach.

References

1. Various Authors, Final Report, DDDAS Workshop at Arlington, VA. (Technical Report, National Science Foundation), 2006
2. Various Authors, Final Report, DDDAS Workshop at Arlington, VA. (Technical Report, National Science Foundation), 2000
3. F. Darema, in *Engineering/Scientific and Commercial Applications: Differences, Similarities, and Future Evolution*, Proceedings of the 2nd Hellenic European Conference on Mathematics and Informatics, vol. 1 (HERMIS, 1994), pp. 367–374
4. J.T. Oden, T. Belytschko, J. Fish, T.J.R. Hughes, C. Johnson, D. Keyes, A. Laub, L. Petzold, D. Srolovitz, S. Yip, in *Simulation-Based Engineering Science: Revolutionizing Engineering Science Through Simulation*. NSF Blue Ribbon Panel on SBES, 2006
5. D. Amsallem, J. Cortial, C. Farhat, Toward real-time CFD-based aeroelastic computations using a database of reduced-order information. AIAA J. **48**, 2029–2037 (2010)
6. J.L. Eftang, D.J. Knezevic, A.T. Patera, An hp certified reduced basis method for parametrized parabolic partial differential equations. Math. Comput. Model. Dyn. Syst. **17**(4), 395–422 (2011)
7. H. Lamari, A. Ammar, P. Cartraud, G. Legrain, F. Jacquemin, F. Chinesta, Routes for efficient computational homogenization of non-linear materials using the proper generalized decomposition. Arch. Comput. Methods Eng. **17**(4), 373–391 (2010)
8. D. Gonzalez, A. Ammar, F. Chinesta, E. Cueto, Recent advances in the use of separated representations. Int. J. Numer. Meth. Eng. **81**(5), 637–659 (2010)
9. F. Chinesta, A. Ammar, E. Cueto, Proper generalized decomposition of multiscale models. Int. J. Numer. Meth. Eng. **83**(8–9), 1114–1132 (2010)
10. A. Ammar, M. Normandin, F. Daim, D. Gonzalez, E. Cueto, F. Chinesta, Non-incremental strategies based on separated representations: applications in computational rheology. Commun. Math. Sci. **8**(3), 671–695 (2010)
11. E. Pruliere, J. Ferec, F. Chinesta, A. Ammar, An efficient reduced simulation of residual stresses in composites forming processes. Int. J. Mater. Form. **3**(2), 1339–1350 (2010)
12. P. Ladeveze, *Nonlinear Computational Structural Mechanics* (Springer, New York, 1999)
13. J.E. Dennis Jr, R.B. Schnabel, *Numerical Methods for Unconstrained Optimization and Nonlinear Equations, Classics in Applied Mathematics* (Corrected Reprint of the 1983 Original), vol. 16 (Society for Industrial and Applied Mathematics, Philadelphia, PA, 1996)
14. A. Ammar, F. Chinesta, P. Diez, A. Huerta, An error estimator for separated representations of highly multidimensional models. Comput. Methods Appl. Mech. Eng. **199**, 1872–1880 (2010)

Chapter 10
Inverse Analysis

This chapter focusses on two different but intimately related problems, frequently appearing in the control of devices or industrial processes, but also in the control of epidemic diseases, for instance. These two problems are the identification of a parameter in a nonlinear dynamical model from on-line measurements, on one hand, and the estimation of a boundary condition at a place where measurements are not possible from redundant information known in other parts of the domain boundary (the so-called Cauchy's problem). This scenario is found when controlling the temperature in the heat exchangers of nuclear plants for example, but it can also appear in many other industrial scenarios. We will consider the one-dimensional Cauchy problem where redundant boundary conditions—namely, temperature and flux—can be measured at one end of the interval, but the required boundary condition for the well-posedness of the problem—temperature at the other end—is not available.

In this chapter a PGD approach to this problem is presented. This approach is based, as the reader can imagine at this level of the book, on the (advantageous) off-line solution of a parametric problem via PGD, for its use on-line in a fast and efficient way. We describe now the main ingredients of such an approach.

10.1 PGD Based Parameter Identification

In this particular application we describe a suitable strategy, based upon PGD techniques, to perform an off-line dynamical system integration procedure that could be then used for on-line control purposes. At this stage, we are not considering the always unavoidable noise related to modeling, control and measurement procedures. Thus, in what follows, measurements correspond to the solution of dynamical systems for a given choice of model parameters, allowing one to circumvent controllability and observability of the inverse techniques issues.

F. Chinesta and E. Cueto, *PGD-Based Modeling of Materials, Structures and Processes*, ESAFORM Bookseries on Material Forming, DOI: 10.1007/978-3-319-06182-5_10,

10.1.1 Dynamic Data-Driven Nonlinear Parametric Dynamical Systems

To describe how PGD could help us in the simulation-based control of processes, we firstly consider a non-linear model usually encountered in the dynamic of populations and therefore in epidemiology modeling. It is well known that, in general, the population size increases in direct relation to its size, but when this size becomes too large its growth rate decreases and sometimes even inverses its tendency. The equation governing this kind of behavior is known as the logistic equation, which reads:

$$\frac{du}{dt} = k \cdot u \cdot (u_\infty - u),$$

where $u = u(t)$, $t \in (0, t_{\max}]$ and whose initial condition is $u(t = 0) = u_g$ with $0 < u_g < u_\infty$.

The exact solution for a constant parameter k is given by:

$$u(t) = \frac{u_\infty}{1 + \left(\frac{u_\infty}{u_g} - 1\right) e^{-k u_\infty t}}.$$

Following standard PGD procedures, parameter k is now assumed unknown, but taking values within a user-prescribed interval $k \in [k_{\min}, k_{\max}]$. The identification of the parameter is a key point in many applications as for example in the simulation and control of epidemic scenarios and also in the on-line adaptation of chemical kinetics governed by similar dynamical models. In order to identify such a parameter, different experimental measurements are carried out at different time instants:

$$\begin{cases} \tilde{t}_1 & \rightarrow \tilde{u}_1 \\ \tilde{t}_2 & \rightarrow \tilde{u}_2 \\ & \vdots \\ \tilde{t}_D & \rightarrow \tilde{u}_D, \end{cases}$$

The main challenge in the simulation and real-time control of the evolution of the field u is how to, based upon experimental measurements, identify as fast as possible the best value of the parameter k, assumed, for simplicity, to be constant within each time interval $(\tilde{t}_i, \tilde{t}_{i+1}]$, $\ i \in 1, \ldots, D - 1$.

In what follows and without loss of generality, we assume that intervals $(\tilde{t}_i, \tilde{t}_{i+1}]$ have the same length, that is, $\tilde{t}_{i+1} - \tilde{t}_i = \Delta$, $i = 1, \ldots, D - 1$. The unknown parameter is assumed taking values in the interval $[k_{\min}, k_{\max}]$ with $k_{\min} > 0$ and $k_{\max} > k_{\min}$. Solving the problem for *any* value of the parameter within this interval provides us with its most general solution, allowing for a complete knowledge of $u(t)$ under all possible circumstances. The PGD approach to this problem, therefore, consists in solving once for life and off-line, the problem: *Find* $u(t, k, u^0)$ *such that*

$$\begin{cases} \frac{du}{dt} = k \cdot u \cdot (u_\infty - u), & t \in (0, \Delta] \\ u(t = 0) = u^0, \end{cases} \tag{10.1}$$

by assuming the initial condition u^0 and the model parameter k as extra-coordinates. Thus one should compute the three-dimensional solution $u(t, k, u^0)$, with $t \in (0, \Delta]$, $k \in [k_{\min}, k_{\max}]$ and $u^0 \in [u_g, u_\infty]$.

Remark 10.1.1 Note that the off-line computation of $u(t, k)$, with $t \in (0, t_{\max}]$, $k \in [k_{\min}, k_{\max}]$, suffices only when the model parameter k is constant in $(0, t_{\max}]$.

If a varying-parameter problem is considered (even if the variation of the parameter is such that k is considered constant within each time interval, for simplicity), an off-line solution is needed that incorporates both the model parameter and the initial condition as extra-coordinates of the model. Thus, the obtained result at time t_i is considered as initial condition for the interval ending at t_{i+1}.

The simplest way of solving Eq. (10.1) consists of defining discrete values of all the model coordinates: $(t_1, t_2, \ldots, t_P)$, $(k_1, k_2, \ldots, k_N)$ and, finally, $(u_1^0, u_2^0, \ldots, u_M^0)$. Without loss of generality we assume such discrete values uniformly distributed, i.e. $t_{i+1} - t_i = h_t, i = 1, \ldots, P-1; k_{i+1} - k_i = h_k$, with $i = 1, \ldots, N-1$ and, finally, $u_{t+1}^0 - u_i^0 = h_u$, with $i = 1, \ldots, M-1$.

Equation (10.1) can be integrated in time by using a finite difference schema (the limitations of such a discretization scheme will be analyzed later on). Again for the sake of simplicity of the exposition, we consider here the simplest strategy, a forward-Euler scheme. Of course, more accurate schemes (particularly, implicit ones) can be considered. Since Eq. (10.1) does not include derivatives with respect to the extra-coordinates (the model parameter and the initial condition), the finite-difference integration of Eq. (10.1) reduces to the solution of the $N \times M$ uncoupled time-dependent ordinary differential equations:

$$\begin{cases} \frac{du^{r,s}}{dt} = k_r \cdot u^{r,s}(t) \cdot (u_\infty - u^{r,s}), & t \in (0, \Delta] \\ u(t = 0) = u_s^0, \end{cases} \tag{10.2}$$

$\forall r, s \in [1, \ldots, N] \times [1, \ldots, M]$, and where $u^{r,s}(t) \equiv u(t; k_r, u_s^0)$. The finite-difference integration of Eq. (10.2) by using a first-order (Euler) explicit schema reads: $\forall r, s, \in [1, \ldots, N] \times [1, \ldots, M]$,

$$\begin{cases} u_1^{r,s} = u_s^0 \\ u_i^{r,s} = u_{i-1}^{r,s} + h_t \cdot k_r \cdot u_{i-1}^{r,s} \cdot (u_\infty - u_{i-1}^{r,s}), & \forall i \in [2, \ldots, P], \end{cases}$$

that allows us to find $u_i^{r,s} = u(t_i, k_r, u_s^0) = u((i-1) \cdot h_t, k_{\min} + (r-1) \cdot h_k, u_g + (s-1) \cdot h_u)$.

The subsequent step in our method will be the parameter identification only from the knowledge of $u_i^{r,s}$. We consider the restriction of $u(t, k, u^0)$ to $u(t = \Delta =$

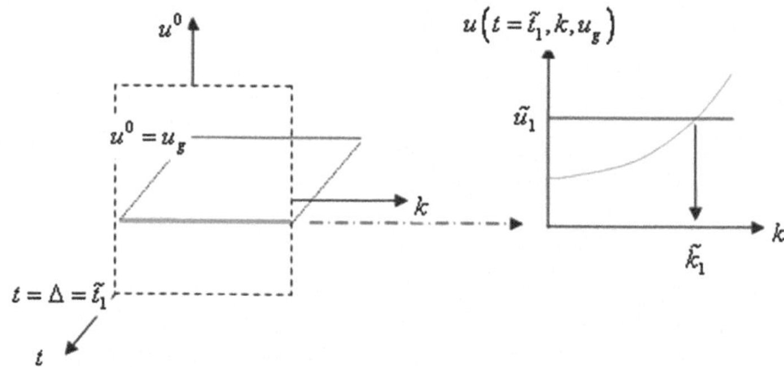

Fig. 10.1 First step of the parameter k identification procedure

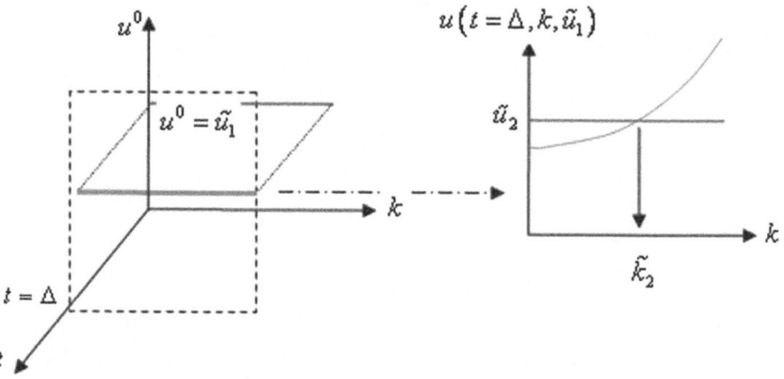

Fig. 10.2 Second step of the parameter identification procedure

$\tilde{t}_1, k, u^0 = u_g)$ and we look for the intersection of such a solution with the experimental value $\tilde{u}_1$. This determines the best-fit value of the parameter k in the interval $(0, \tilde{t}_1]$, $\tilde{k}_1$, as illustrated in Fig. 10.1.

At this point we consider again the restriction of $u(t, k, u^0)$ to $u(t = \Delta, k, u^0 = \tilde{u}_1)$ and we look for the intersection of this solution with the experimental value $\tilde{u}_2$. This point determines the best-fitting value for the parameter k in the interval $[\tilde{t}_1, \tilde{t}_2]$, $\tilde{k}_2$, as shown in Fig. 10.2.

It is noteworthy to highlight that the integration in the whole time domain is performed from the only knowledge of $u_i^{r,s}$, despite the eventual variability of the model parameter. This solution is computed once and off-line. Therefore, we do not care too much about the CPU time involved in this calculation. Since the initial condition u_0 has been introduced as an extra-dimension of the problem, the integration can be adapted dynamically on-line (at extremely high feedback rates) to experimental data coming from sensors, by adapting the value of the model parameter necessary to fit the experimental value. This value therefore is considered as initial condition for the integration in the next time interval Δ.

It has been already mentioned that, if instead of computing $u(t, k, u^0)$, with $t \in (0, \Delta]$, one computes $u(t, k)$ with $t \in (0, t_{\max}]$, from the given initial condition u_g, the value of k for the whole process could be identified as soon as an experimental data is available. However, taking into account that the change of the model parameter is not possible, and it implies the on-line integration of the model $u(t, k)$ at each interval $t \in (\tilde{t}_i, \tilde{t}_{i+1})$ taking the available data $\tilde{u}_i$ as initial condition. The drawback of such a procedure is the need to perform on-line integrations, a fact that greatly limits its applicability in real-time control and therefore in DDDAS.

Remark 10.1.2 Under the weak assumption that the time step between two successive measurements Δ is constant we can easily enforce the integration time step P to be coincident with Δ. Notice, however, the data $\tilde{u}_i$ rarely coincides with a value of u_s^0. To solve this, different possibilities exist:

1. Consider the nearest u_s^0 to $\tilde{u}_i$;
2. Consider the two neighboring values u_{s+1}^0 and u_s^0. This allows to define the interval to which $\tilde{k}_{i+1}$ belongs
3. Consider the interpolated solution from $u(t = \Delta, k, u_{s+1}^0)$ and $u(t = \Delta, k, u_s^0)$.

In what follows we are analyze the last alternative.

10.1.2 Proper Generalized Decomposition of Dynamical Systems

At this point, the reader has been undoubtedly able to imagine the strategy suggested in this book to solve this class of simulation-based control problems governed by dynamical data. Of course, PGD allows for a very efficient solution within the framework stated in previous sections. Here, we detail the necessary ingredients to accomplish it.

The method introduced above is obviously applicable by using any discretization technique, such as finite differences, for instance. Indeed, it should work very well for the example shown in the next section. But if the number of state dimensions of the model (parameters, boundary conditions, etc., as explained before) increases, the only way to deal with them seems to be the PGD method here presented. For the sake of simplicity we keep, however, the number of dimensions of the problem restricted to three, without loss of generality:

$$\begin{cases} \frac{du}{dt} = k \cdot u \cdot (u_\infty - u) \\ u(t = 0) = u^0, \end{cases} \tag{10.3}$$

with $t \in I = (0, \Delta]$, $k \in \aleph = [k_{\min}, k_{\max}]$ and $u^0 \subset \Im = [u_g, u_\infty]$.

As mentioned before, we intend to introduce parameter k as well as the initial condition u^0 as additional coordinates in the problem. Since PGD works naturally with homogeneous boundary and initial boundary conditions, to introduce the initial condition in the ODE a simple change of variables is suggested:

$$\hat{u} = u - u^0,$$

which, once introduced in Eq. (10.3) gives rise to

$$\begin{cases} \frac{d\hat{u}}{dt} = k(\hat{u} + u^0)(u_\infty - \hat{u} - u^0), & t \in (0, \Delta] \\ \hat{u}(t = 0) = 0. \end{cases} \tag{10.4}$$

At this point, the reader is already familiar with the fundamental characteristic of PGD, i.e., that it seeks the solution to Eq. (10.4) in the form:

$$\hat{u} \approx \sum_{i=1}^{Q} T_i(t) \cdot K_i(k) \cdot U_i(u^0). \tag{10.5}$$

As already mentioned throughout this book, proper generalized decomposition constitutes an *a priori* method, i.e., it does not assume any particular form of the functions T_i, K_i or U_i, these must be generated during the execution of the method itself. This gives rise to a nonlinear problem that is described next. Within the iterative process inherent to nonlinear problems, we assume that the approximation at iteration $n < Q$ is already achieved,

$$\hat{u}^n = \sum_{i=1}^{n} T_i(t) \cdot K_i(k) \cdot U_i(u^0),$$

such that in the current iteration the functional product $T_{n+1}(t) \cdot K_{n+1}(k) \cdot U_{n+1}(u^0)$ is sought. With an eye towards the simplification of the notation, we will denote this product as $T_{n+1}(t) \cdot K_{n+1}(k) \cdot U_{n+1}(u^0) = R(t) \cdot S(k) \cdot W(u^0)$. Prior to solving this nonlinear problem, a linearisation is mandatory. The simplest choice consists in a fixed-point iterative scheme along alternated directions. This scheme assumes $S(k)$ and $W(u^0)$ known and proceeds through the determination of $R(t)$. With this function $R(t)$ just computed and with the previous $W(u^0)$ it proceeds by calculating $S(k)$ and, finally, with the just obtained $R(t)$ and $S(k)$, it determines $W(u^0)$. This procedure continues until convergence. Functions thus computed constitute the next term in the approximation, $T_{n+1}(t) = R(t)$, $K_{n+1}(k) = S(k)$ and $U_{n+1} = W(u^0)$.

All the mentioned steps are detailed in what follows.

10.1.2.1 Computation of $R(t)$ from $S(k)$ and $W(u^0)$

Consider now the linearized weak form of the problem (10.4) that consists of considering all the non-linear terms at the previous iteration, i.e. $\hat{u}^n$:

$$\int_{I \times \aleph \times \Im} \hat{u}^* \left(\frac{\partial \hat{u}}{\partial t} - k(\hat{u}^n + u^0)(u_\infty - u^0 - \hat{u}) \right) dt\, dk\, du^0 = 0, \tag{10.6}$$

in which the trial and test functions, respectively, are

$$\hat{u}(t, k, u^0) = \sum_{i=1}^{n} T_i(t) \cdot K_i(k) \cdot U_i(u^0) + R(t) \cdot S(k) \cdot W(u^0) \tag{10.7}$$

and

$$\hat{u}^*(t, k, u^0) = R^*(t) \cdot S(k) \cdot W(u^0). \tag{10.8}$$

Introducing (10.7) and (10.8) in the weak form (10.6) we obtain

$$\begin{aligned}
&\int_{I\times\aleph\times\Im} R^* \cdot S \cdot W \cdot \left(\frac{dR}{dt} \cdot S \cdot W + k \cdot \left(\sum_{i=1}^{n} T_i \cdot K_i \cdot U_i + u^0 \right) R \cdot S \cdot W \right) dt\, dk\, du^0 \\
&= - \int_{I\times\aleph\times\Im} R^* \cdot S \cdot W \left(\sum_{i=1}^{n} \frac{dT_i}{dt} \cdot K_i \cdot U_i \right. \\
&\quad \left. + k \left(\sum_{i=1}^{n} T_i \cdot K_i \cdot U_i + u^0 \right) \left(\sum_{i=1}^{n} T_i \cdot K_i \cdot U_i - u_\infty + u^0 \right) \right) dt\, dk\, du^0,
\end{aligned} \tag{10.9}$$

Since all the functions depending on the parametric coordinate k and the initial condition u^0 are known, it is possible to perform the numerical integration of these functions along their respective domains $\aleph \times \Im$.

Equation (10.9), after integration in $\aleph \times \Im$, represents the weak form of an ODE that defines the temporal evolution of the field $R(t)$. It could be solved by any available discretization technique (SU, discontinuous Galerkin, ...). The strong form of Eq. (10.9), after integrating it in $\aleph \times \Im$, will be of the type

$$w_1 \cdot s_1 \cdot \frac{dR(t)}{dt} + w_2 \cdot s_2 \cdot R(t) + \left(\sum_{i=1}^{n} \alpha_i \beta_i T_i(t) \right) R(t) = f(t), \tag{10.10}$$

with

$$\begin{cases}
w_1 = \int_{\Im} W^2 du^0 \\
w_2 = \int_{\Im} u^0 W^2 du^0 \\
s_1 = \int_{\aleph} S^2 dk \\
s_2 = \int_{\aleph} k \cdot S^2 dk \\
\alpha_i = \int_{\Im} W^2 U_i du^0 \\
\beta_i = \int_{\aleph} S^2 K_i k dk
\end{cases} \tag{10.11}$$

and $f(t)$ the function that results from integrating the right-hand member in the $\Im$ and $\aleph$ intervals.

10.1.2.2 Computation of $S(k)$ from $R(t)$ and $W(u^0)$

In this case, the weighting function will be

$$\hat{u}^*(t,k,u^0) = S^*(k)\cdot R(t)\cdot W(u^0). \tag{10.12}$$

This gives rise to a weak form

$$\begin{aligned}\int_{I\times\aleph\times\Im} & S^*\cdot R\cdot W\cdot\left(\frac{dR}{dt}\cdot S\cdot W + k\cdot\left(\sum_{i=1}^{n} T_i\cdot K_i\cdot U_i + u^0\right)R\cdot S\cdot W\right)dt\,dk\,du^0\\ &= -\int_{I\times\aleph\times\Im} S^*\cdot R\cdot W\cdot\left(\sum_{i=1}^{n}\frac{dT_i}{dt}\cdot K_i\cdot U_i\right.\\ &\quad\left. + k\left(\sum_{i=1}^{n} T_i\cdot K_i\cdot U_i + u^0\right)\left(\sum_{i=1}^{n} T_i\cdot K_i\cdot U_i - u_\infty + u^0\right)\right)dt\,dk\,du^0,\end{aligned} \tag{10.13}$$

The neat difference between the problem defined by Eq. (10.13) and that in Eq. (10.9) is that now no differential operator is involved. Thus the strong form resulting from the problem reads:

$$r_2\cdot w_1\cdot S(k) + r_1\cdot w_2\cdot k\cdot S(k) + \left(\sum_{i=1}^{n}\gamma_i\alpha_i K_i(k)\right)kS(k) = g(k), \tag{10.14}$$

with

$$\begin{cases} r_1 = \int_I R^2 dt\\ r_2 = \int_I R\cdot\frac{dR}{dt}\\ \gamma_i = \int_I T_i R^2 dt\end{cases} \tag{10.15}$$

and $g(k)$ the function that results after integration of the right-hand member in the I and $\Im$ intervals. This represents an algebraic equation. Note that the introduction of parameters as additional spatial coordinates in the problem does not have a relevant effect in the computational cost of the resulting PGD approximation but it could induce the necessity of considering much more terms in the finite sums decomposition.

10.1.2.3 Computation of $W(u^0)$ from $R(t)$ and $S(k)$

In this last case the weighting function results:

$$\hat{u}(t,k,u^0) = W^*(u^0)\cdot R(t)\cdot S(k). \tag{10.16}$$

and therefore we arrive at a problem whose weak form reads

$$\int_{I\times\aleph\times\Im} W^* \cdot R \cdot S \cdot \left(\frac{dR}{dt} \cdot S \cdot W + k \left(\sum_{i=1}^{n} T_i \cdot K_i \cdot U_i + u^0 \right) \cdot R \cdot S \cdot W \right) dt\, dk\, du^0$$
$$= - \int_{I\times\aleph\times\Im} W^* \cdot R \cdot S \cdot \left(\sum_{i=1}^{n} \frac{dT_i}{dt} \cdot K_i \cdot U_i \right.$$
$$\left. + k \left(\sum_{i=1}^{n} T_i \cdot K_i \cdot U_i + u^0 \right) \left(\sum_{i=1}^{n} T_i \cdot K_i \cdot U_i - u_\infty + u^0 \right) \right) dt\, dk\, du^0 \tag{10.17}$$

again, Eq. (10.17) does not involve any differential operator. Its associated strong form reads

$$(r_2 \cdot s_1 + r_1 \cdot s_2 \cdot u^0) \cdot W(u^0) + \left(\sum_{i=1}^{n} \gamma_i \beta_i U_i(u^0) \right) W(u^0) = h(u^0), \tag{10.18}$$

with $h(u^0)$ the function that results after integration of the right hand member in the I and $\aleph$ intervals. This results, again, in an algebraic equation.

10.2 PGD Based Solution of Cauchy's Problem

An intimately related problem to that of on-line inverse identification of parameters frequently arises in industrial control settings. The problem consists of estimating unknown boundary conditions (operating conditions of some devices at places where measurements can not be performed). To provide the reader with a clear picture of this type of problems in industrial settings, we consider the numerical solution of the transient heat (Fourier) equation in a one-dimensional domain, where redundant boundary conditions, temperature and thermal flux, are known at one boundary of the domain, but where both fields are unknown at the other boundary. This problem is frequently encountered in the control of thermal exchangers. In such a device, it is not possible to fix thermocouples at the internal surface of a pipe (e.g. exchangers in nuclear plants) and the temperature of the fluid flowing inside must be inferred from redundant information known in the external surface of the pipe, where temperature and heat flux can be experimentally measured.

For control purposes it is necessary to provide the system with an accurate estimate of the unknown filed (here, temperature) and under real-time feedback constrains. These depend strongly on the particular pursued application. In this example, it is necessary to identify temperature fluctuations occurring suddenly in order to take the pertinent control decisions.

This problem settings define an inverse problem that can be solved by applying a variety of well-established techniques. A naive approach, for instance, could imagine that at each time step an arbitrary trial temperature could be enforced at the internal (computational) boundary, then solving the heat equation for both boundary temperatures, the one really measured and the estimated one, and then computing the heat flux in the boundary where it is known. By comparing the computed and experimentally measured heat fluxes the unknown boundary temperature could be updated by using an appropriate technique trying to minimize the difference between both fluxes. As soon as the unknown temperature is updated the thermal problem can be solved again and the procedure repeated until reaching the convergence, i.e. until obtaining a gap lower than a small enough value. After reaching convergence, one can move to the next time step in which the just described iteration procedure should be repeated. Other more sophisticated strategies exist, and even if there are no major conceptual difficulties, the real time constraint is not easy to address.

In what follows we exemplify again how to employ PGD approaches to this problem. It will follow similar guidelines to the problems presented in previous sections. We suggest particularly a procedure able to address the inverse identification problem accurately and in real time. For the sake of simplicity we consider a linear thermal model defined in a one-dimensional domain. The described procedure, however, can be easily extended for considering more realistic scenarios: nonlinear constitutive equations, two or three dimensional geometries, to name but a few examples.

Consider the linear transient heat equation

$$\frac{\partial u}{\partial t} - k\frac{\partial^2 u}{\partial x^2} = 0 \text{ in } (0, L) \times (0, t_{\max}]$$

with boundary conditions

$$\begin{cases} u(x,0) = u^0(x) \\ u(0,t) = u_g(t) \\ \frac{\partial u}{\partial x}(0,t) = q_g(t) \end{cases} .$$

Do not forget that the aim of the proposed method is actually to perform inverse identification of boundary conditions. Given a set of temperature values $\left\{u_g^i\right\}_{i=1}^D$ and heat fluxes $\left\{q_g^i\right\}_{i=1}^D$, coming from experimental measurements taken at the left boundary of the domain, say $x = 0$, one must estimate the time evolution of the temperature at the right boundary, say $x = L$, denoted by $\theta(t)$, at least at the same time instants $\left\{\theta^i\right\}_{i=1}^D$. We assume again that measurements are performed at each time interval Δ.

Again we exploit the fact that Proper Generalized Decomposition allows us to compute a multidimensional solution of the thermal model, where initial and boundary conditions could efficiently be considered as state-space extra-coordinates.

If the initial condition is parametrized by a piece-wise linear finite element approximation space, defined from p nodes uniformly distributed in the interval $[0, L]$, x_i, $i = 1, \ldots, p$, the separated representation of the unknown temperature field reads:

$$u\left(x, t, u_g, u_2^0, \ldots, u_{p-1}^0, \theta\right) \approx \sum_{i=1}^{i=N} X_i(x) \cdot T_i(t) \cdot U_i^1(u_g) \cdot$$
$$U_i^2(u_2^0) \cdot \ldots \cdot U_i^{p-1}(u_{p-1}^0) \cdot U_i^p(\theta),$$

where $x \in (0, L)$, $t \in (0, \Delta]$, $u_i^0 \in [u_{min}, u_{max}]$, $i = 2, \ldots, p-1$, $u_g \in [u_{min}, u_{max}]$ and $\theta \in [u_{min}, u_{max}]$.

This separated representation can be constructed by following the procedure previously described in this chapter. Once obtained, space derivatives involved in the heat flux can be computed by

$$\frac{\partial u\left(x, t, u_g, u_2^0, \ldots, u_{p-1}^0, \theta\right)}{\partial x} \approx \sum_{i=1}^{i=N} \frac{dX_i(x)}{dx} \cdot T_i(t) \cdot U_i^1(u_g) \cdot$$
$$U_i^2(u_2^0) \ldots U_i^{p-1}(u_{p-1}^0) \cdot U_i^p(\theta).$$

Assuming known the temperature field at time $t = n \cdot \Delta$, at the subsequent time step $n + 1$, the temperature θ^{n+1} must be computed from the information known at $x = 0$, u_g^{n+1} and q_g^{n+1}, and that temperature field at the previous time step. Let us therefore consider:

$$\left.\frac{\partial u\left(x, t, u_g, u_2^0, \ldots, u_{p-1}^0, \theta\right)}{\partial x}\right|_{\substack{x=0, t=\Delta, u_g=u_g^{n+1}, u_2^0=u^n(x_2), \ldots, \\ u_{p-1}^0=u^n(x_{p-1}), \theta=\theta^{n+1}}} = q_g^{n+1},$$

that represents a single equation to determine the single unknown variable θ^{n+1}. In this equation, $u^n(x_k)$, $k = 2, \ldots, p-1$ comes from the final values at the previous time step according to:

$$u^n(x_k) = u(x = x_k, t = \Delta, u_g = u_g^n, u_2^0 = u^{n-1}(x_2), \ldots,$$
$$u_{p-1}^0 = u^{n-1}(x_{p-1}), \theta = \theta^n),$$

where similar expressions apply for $u^{n-1}(x_k)$, $k = 2, \ldots, p-1$.

Note that, here, an important assumption has been made. Notice that it has been assumed that the boundary conditions u_g^i and q_g^i are constant within the interval $[(i-1) \cdot \Delta, i \cdot \Delta]$, $i = 2, \ldots, D$. Other possibilities exist, of course, the simplest one being a linear evolution between each two consecutive measurements. In that case, the separated representation of the temperature field reads $u\left(x, t, u_g, u_1^0, u_2^0, \ldots, u_{p-1}^0, u_p^0, \theta\right)$, where $x \in (0, L)$, $t \in (0, \Delta]$, $u_i^0 \in [u_{min}, u_{max}]$, $i = 1, \ldots, p$, $u_g \in [u_{min}, u_{max}]$ and $\theta \in [u_{min}, u_{max}]$.

Once the multidimensional separated representation has been constructed, the identification procedure follows these steps:

$$\left.\frac{\partial u\left(x, t, u_g, u_1^0, u_2^0, \ldots, u_{p-1}^0, u_p^0, \theta\right)}{\partial x}\right|_{\substack{x=0, t=\Delta, u_g=u_g^{n+1}, u_1^0=u_g^n, u_2^0=u^n(x_2), \ldots, \\ u_{p-1}^0=u^n(x_{p-1}), u_p^0=\theta^n, \theta=\theta^{n+1}}} = q_g^{n+1},$$

that represents again a single equation to compute the single unknown θ^{n+1}.

10.3 Numerical Example: Dynamic Data Driven Nonlinear Parameter Identification

To illustrate the just described procedure we consider now the dynamical model given by Equation in Sect. 10.1.1 where $u_\infty = 1$, $u_g = 0.2$, $t \in (0, 10]$ and the unknown model parameter taking values in the interval $k \in [0, 3]$.

In this example, we assume that the field u is measured experimentally at times $\tilde{t}_i = i \cdot \Delta$, $i = 1, \ldots, 10$ and $\Delta = 1$. Linear finite element meshes were chosen, with 200, 60 and 160 elements, respectively, along each coordinate. We consider two scenarios.

10.3.1 A First Scenario

We consider firstly experimental measurements assumed exact, i.e., coming from the exact solution of Equation in Sect. 10.1.1 with $k = 0.7$ (the model parameter is assumed constant in the whole simulation time interval (0, 10]). In that case the measured values at time instants $\tilde{t} = 1, 2, \ldots, 10$ are given by $\tilde{u} = 0.335, 0.503, 0.671, 0.804, 0.892, 0.943, 0.971, 0.985, 0.993, 0.996$.

Figure 10.3 depicts the restriction of $u(t, k, u^0)$ at time $t = \Delta$, i.e., $u(t = \Delta, k, u^0)$. Notice (Fig. 10.3, right) the good accuracy obtained with the PGD method by employing only one product of functions. This error ($\mathscr{L}_2$-norm) is $1.45 \cdot 10^{-4}$ for the mentioned single product of functions, shown in Fig. 10.4.

Remark 10.3.1 It is well known in the literature that the forward Euler, finite difference discretization of the Verhulst logistic Equation in Sect. 10.1.1 gives rise to a recurrence relation very much like the logistic map. As it is well known, the logistic map presentschaotic behavior for values of the product $k \cdot \Delta$ bigger than 3.57,

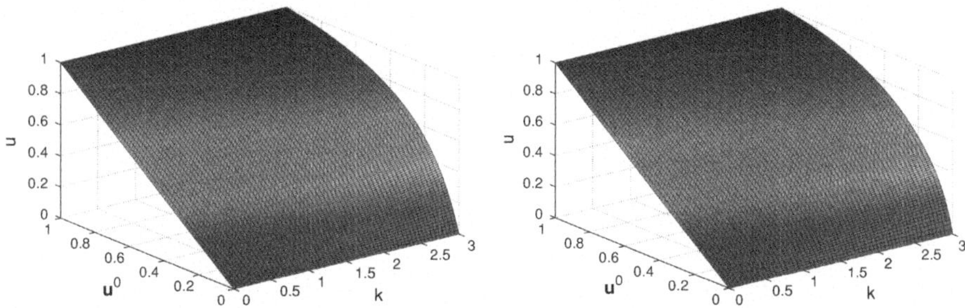

Fig. 10.3 Solution $u(t = \Delta, k, u^0)$. *Left*, exact solution. *Right*, numerical one obtained by applying PGD

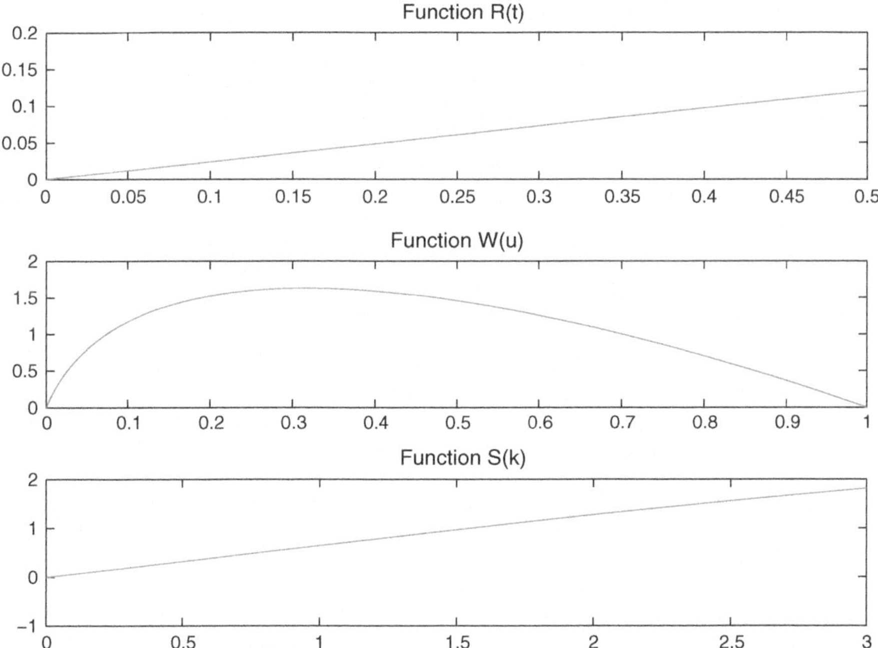

Fig. 10.4 Functions employed for the numerical solution of the problem

showing oscillations dependent on the initial condition for $k > 2$. In the simulations that follow, chaotic or oscillatory behavior is avoided by a judicious choice of time discretization.

Figure 10.5 depicts the restriction of solution $u(t = \Delta, k, u_0)$ shown in Fig. 10.3 to the given initial condition $u^0 = u_g = 0.2$, that is: $u(t = \Delta, k, u^0 = 0.2)$. Since such value $u^0 = u_g$ does not coincide with a discrete value of the initial condition axis we consider the interpolated curve $u(t = \Delta, k, u_g) = \sum_{i=1}^{n} T_i(\Delta) \cdot K_i(k) \cdot U_i(u_g)$.

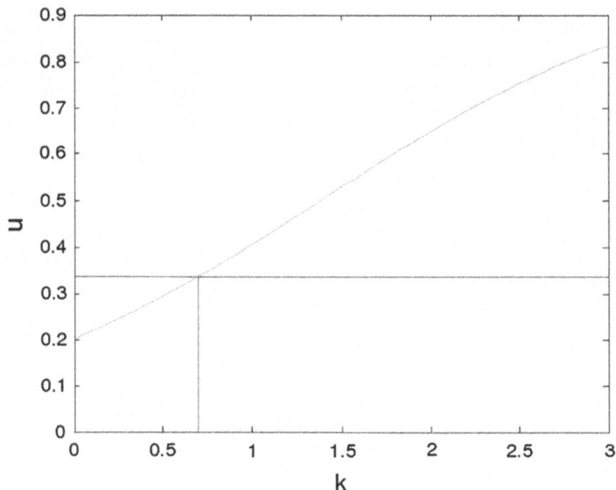

Fig. 10.5 Solution $u(k) = u(t = \Delta, k, u^0 = 0.2)$

Table 10.1 Parameter estimation for $k^{\text{exact}} = 0.7$

$\tilde{t}$	$\tilde{u}$	k^{exact}	$\tilde{k}_{\text{ident}}$
1	0.335	0.7	0.701
2	0.503	0.7	0.700
3	0.671	0.7	0.699
4	0.804	0.7	0.699
5	0.892	0.7	0.699
6	0.943	0.7	0.699
7	0.871	0.7	0.699
8	0.985	0.7	0.699
9	0.993	0.7	0.699
10	0.996	0.7	0.699

The resulting interpolated solution is depicted in cyan in Fig. 10.5, whereas the reference solution, $\tilde{u}_1 = 0.335$ is depicted in red. The intersection point of both curves defines the optimal model parameter k, $k = 0.701$, that in this case is very close to the exact one $k = 0.7$ that served to construct the set of sample measurements. This procedure continues as described in the previous section. Table 10.1 gives an overview of the resulting identified parameters.

10.3.2 Random Variation of the Parameter

In what follows a second, slightly different, scenario is considered. Now, parameter k is defined randomly in each time interval, according to the expression: $(\tilde{t}_i, \tilde{t}_{i+1})$, $i = 1, \ldots, D = 10$, with

Table 10.2 Parameter estimation for a parameter evolving randomly

$\tilde{t}$	$\tilde{u}$	k^{exact}	$\tilde{k}_{\text{ident}}$
1	0.742	2.444	2.445
2	0.978	2.717	2.703
3	0.984	0.381	0.381
4	0.999	2.740	2.722
5	0.999	1.897	1.893
6	0.999	0.293	0.294
7	0.999	0.835	0.838
8	0.999	1.641	1.639
9	0.999	2.873	2.858
10	0.999	2.895	2.880

$$k^{\text{exact}} = \varepsilon \cdot \mathscr{U},$$

where $\varepsilon = 3$ and $\mathscr{U}$ denotes a uniform random variable defined in the interval $[0, 1]$. Table 10.2 shows the identified parameters. We can notice the high accuracy obtained by employing the on-line parameter identification procedure previously described. Note that it only needed a single off-line solution of a model defined in a high dimensional space, that was efficiently solved by using the Proper Generalized Decomposition.

10.4 Numerical Example: Dynamic Data Driven Cauchy Problem

In order to evaluate the reliability of the strategy introduced in Sect. 10.1, two test cases are considered. For the first one, the temperature on the right boundary evolves linearly in time whereas in the second scenario a discontinuous step is considered.

10.4.1 Identifying a Linear Evolution on the Unknown Boundary Condition

Let us assume firstly an initial constant value of the temperature, followed by a linear evolution in time at $x = L$, that constitutes the sought boundary condition, whereas at $x = 0$ a convection-type boundary condition is considered that allows us to solve the equation by using a finite-difference technique. Once obtained, the temperature derivative is computed at $x = 0$. The values of the temperature and its derivatives at $x = 0$ each $\Delta = 1$ are retained while we try to identify the profile at $x = L$. Firstly, it is assumed that the boundary conditions u_g^{n+1} and q_g^{n+1} are constant within the interval $[n \cdot \Delta, (n + 1) \cdot \Delta]$, $n = 1, \ldots, D - 1$. In the numerical experiments that follow the following values have been considered: $k = 0.1$ and $L = 1$.

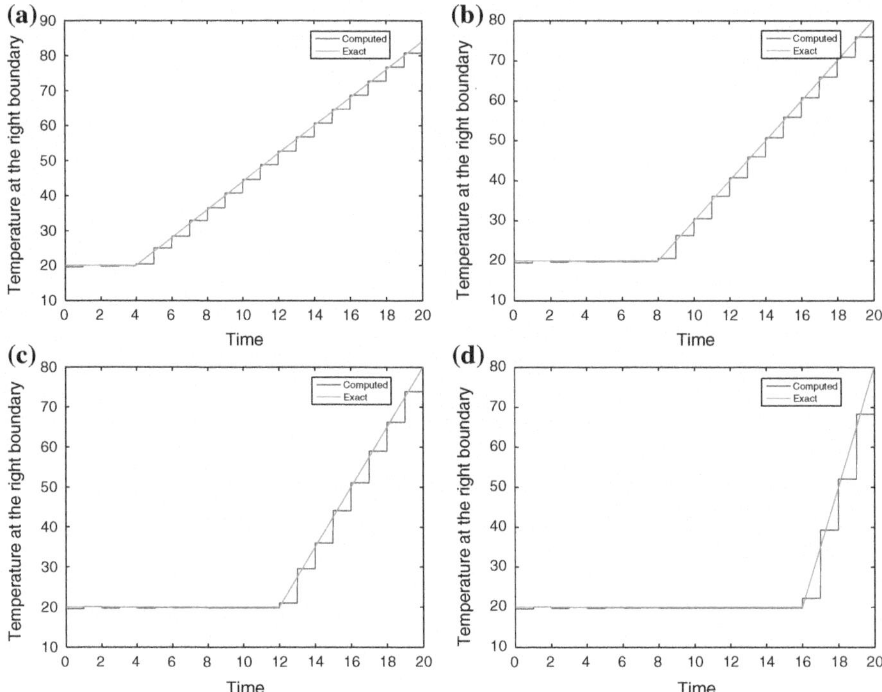

Fig. 10.6 Identified solution for different rates of increase of the identified temperature. **a** slope = 4, **b** slope = 5, **c** slope = 7.5, **d** slope = 15

At each Δ the temperature on the right boundary, θ, is identified as previously described. As soon as this temperature is calculated the temperature for $t \in (t_n, t_{n+1}] = (n \cdot \Delta, (n+1) \cdot \Delta]$ reads

$$u\left(x, t, u_g^{n+1}, u^n(x_2), \ldots, u^n(x_{p-1}), \theta^{n+1}\right). \tag{10.19}$$

Obviously, if we particularize it on the right-hand boundary we obtain:

$$u\left(x = L, t, u_g^{n+1}, u^n(x_2), \ldots, u^n(x_{p-1}), \theta^{n+1}\right) = \theta^{n+1}, \quad \forall t \in [t_n, t_{n+1}]. \tag{10.20}$$

Thus one can expect a discontinuity of the identified temperature on the right-hand boundary at each sampling time t_n, as noticed in all the solutions depicted in subsequent figures.

Figure 10.6 depicts the identified evolution at $x = L$, that is compared with the exact one. The impact of the hypothesis of constant boundary conditions within each time interval Δ can be noticed for different values of the rate of increase of the temperature at $x = L$. The identified temperature follows the exact evolution

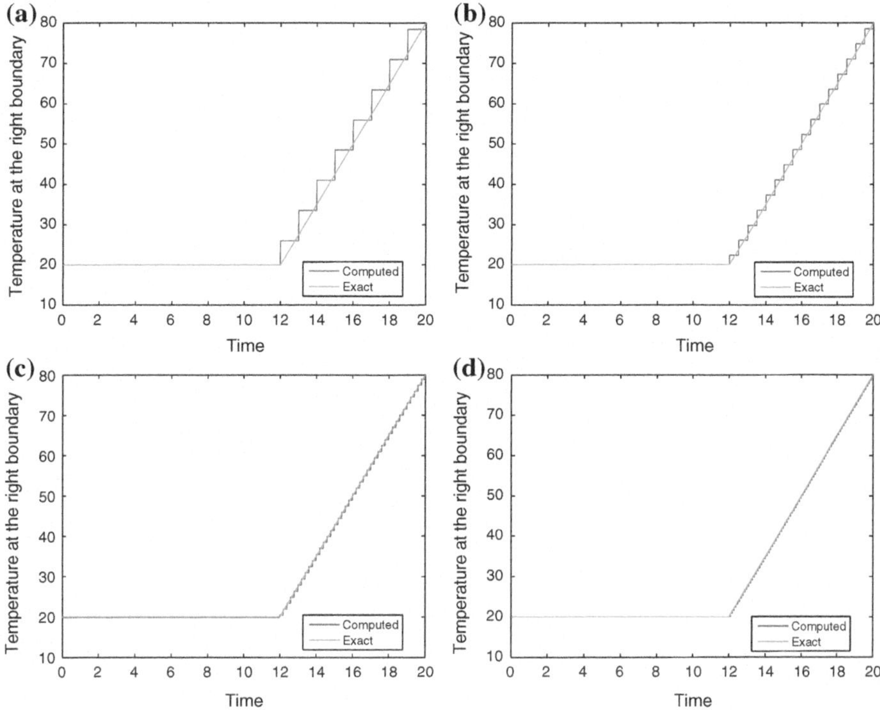

Fig. 10.7 Influence of the sampling frequency. **a** f = 1 Hz, **b** f = 2 Hz, **c** f = 5 Hz, **d** f = 10 Hz

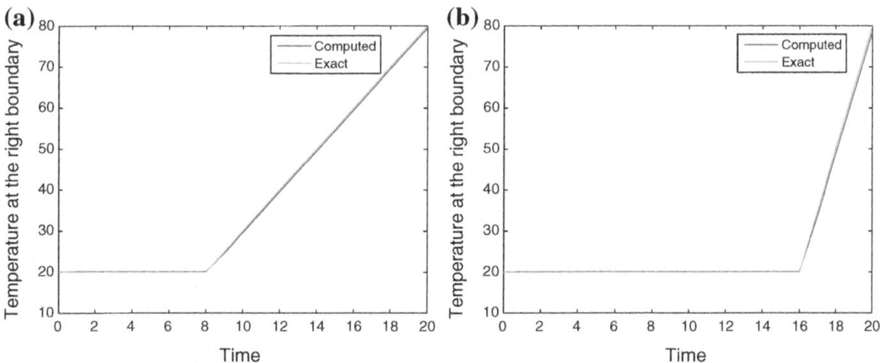

Fig. 10.8 Considering a linear evolution of the boundary conditions between two consecutive sampling times. **a** slope = 5, **b** slope = 15

but it underestimates the temperature values. Obviously, by increasing the sampling frequency the accuracy is notably increased, as proved in Fig. 10.7. The just cited issues disappear as soon as a piecewise linear evolution of the boundary conditions is assumed. This is the case illustrated in Fig. 10.8.

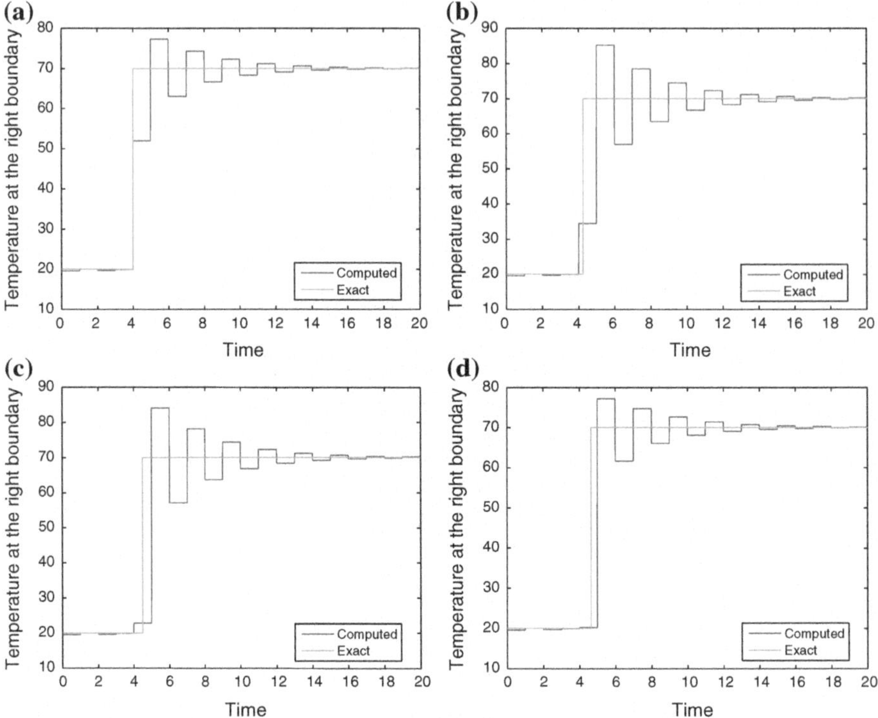

Fig. 10.9 Identified solution for different positions of the discontinuity within the interval $T = \Delta$. **a** 0, **b** $\frac{T}{a}$, **c** $\frac{T}{4}$, **d** $\frac{3T}{4}$

10.4.2 Identifying Discontinuous Evolutions in the Unknown Boundary Condition

When considering a discontinuity in the temperature profile at the right-hand boundary the accuracy of the identified temperature evolution could depend on several factors: (i) the material conductivity that controls the delay between a sudden increase of the temperature and its effects on the opposite boundary, (ii) the sampling frequency and (iii) location of the discontinuity within the interval $T = \Delta$.

When considering a low diffusion coefficient the identification is quite poor because of delay effects in transfer of the information throughout the domain. The accuracy cannot be significantly improved by modifying the location of the discontinuity within the interval $T = \Delta$ as illustrated in Fig. 10.9 for $k = 0.1$.

When increasing the thermal conductivity the identification is significantly improved and as expected the best identification is carried out when the discontinuity is located exactly in a sampling instant as shown in Fig. 10.10.

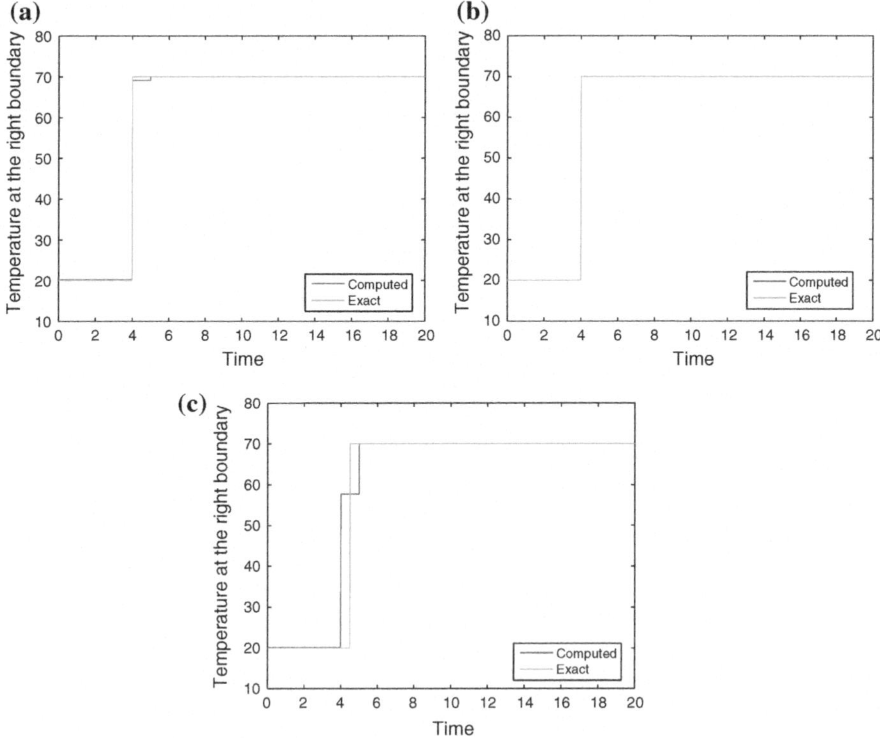

Fig. 10.10 Influence of the conductivity and location of the discontinuity. **a** k = 0.5/0, **b** k = 1/0, **c** k = $0.5/\frac{T}{2}$

10.5 Conclusions

In this chapter a method for efficient simulation in the context of DDDAS based upon the use of Proper Generalized Decompositions has been presented and its capabilities analyzed by means of suitable, yet academic, examples. The method is based upon the construction of a high dimensional solution for any value of the parameters considered as state space coordinates. In this chapter two different, but frequently related problems have been considered. These problems were the inverse parameter identification in a problem governed by nonlinear dynamic equations and dynamic data, and a second one consisting in estimating boundary conditions at places where measurements are not possible (inverse identification of boundary conditions). PGD techniques allow us to post-process the solution in order to obtain on-line an accurate and fast solution of the problem, rather than simulating the evolution of the problem for any change in the parameters.

Chapter 11
Virtual Design: Automated Tape Placement

Production of large pieces made of thermoplastic composites like plates or shells is a challenging issue for nowadays industry. Thermoplastic composites are composites that use a thermoplastic polymer as a matrix. This category still represents a niche market because of the difficulties associated to its processing. Several reliable manufacturing processes are now available for building up thermoplastic laminated structures. Among them, the automated tape placement (ATP) appears to be an appealing strategy. In this process a tape is placed and progressively welded on the substrate. By laying additional layers in different directions, a part with desired properties and geometry can be produced. However, welding of two thermoplastic pieces requires specific physical conditions: a permanent contact, also called intimate contact, and a temperature that has to be high enough to allow the diffusion of macromolecules. Due to the low thermal conductivity of thermoplastic composites, a high temperature at the interface can only be reached with local heating. ATP uses a laser (or sometimes hot gas torches) and a cylindrical consolidation roller to ensure both conditions required for a proper welding, as depicted on Fig. 11.1.

Numerical simulations of such a process is currently a subject of intensive research work. Indeed, because of the successive heating and cooling of the structure during the addition of new tapes, residual stresses develop in the formed part. The evaluation of this residual stress is crucial because it has a significant impact on both the mechanical properties and the geometry of the manufactured plate or shell. It can in particular lead to a distortion of the component, inter-ply delaminations or matrix cracking. High-levels of stress may arise because of two factors. First, the large discrepancies in the thermal expansion coefficients of matrix and fiber materials lead to important deformation gradients at the matrix/fiber interface. Another source of stress is the gradient of temperature inside the laminated structure. Because two neighboring plies do not have necessarily the same fiber orientation their thermal expansion behavior tends to be incompatible.

Experimentally, it is quite difficult to measure residual stresses. Destructive methods use the release of stress and its associated strain when performing a cutting of the structure. Non-destructive methods like X-ray diffraction or neutron diffraction

F. Chinesta and E. Cueto, *PGD-Based Modeling of Materials, Structures and Processes*, ESAFORM Bookseries on Material Forming, DOI: 10.1007/978-3-319-06182-5_11,

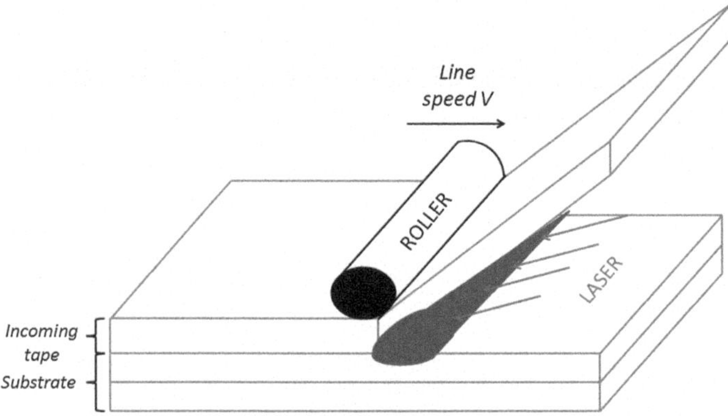

Fig. 11.1 Sketch of automated tape placement

are more accurate but still very expensive. Numerical simulation turns out therefore to be one of the cheapest and most promising alternative but several issues related to the process itself make the task quite complicated as we are going to expose throughout this work

Several models have been proposed since the early 1990's. We can mention in particular the numerical analysis made by Sonnez and al. [1] and the work by Pitchumani and al. who were interested in the study of interfacial bonding [2]. In this latter, the domain considered is only 2D. Regarding the thermal model, it consists of numerically solving the transient heat transfer equation. However, strong assumptions are made concerning the boundary conditions. Indeed, no heat transfer by conduction is considered at the interface with the mandrel and in the region of contact with the roller. Moreover, in order to simplify the geometry of the domain, an incoming tow is assumed to be instantaneously laid down all along the substrate, which is far from being the case in the real process. And finally, the contact is assumed to be perfect at the interface of 2 plies which seems to be also a relatively strong assumption.

Here we propose a different strategy based on a slightly more complex model. First of all, the domain we consider is 3D and the material anisotropic. In order to take into account the imperfect adhesion at the inter-ply interface, thermal contact resistances are introduced. Regarding the mechanical problem, the incoming tow is progressively laid down on the substrate and is subjected to a tension force in order to reproduce the pre-tension applied in the real process. But actually, beyond the model itself, the numerical method employed for the solution of the thermal and mechanical problems associated to the ATP process is novel. To our knowledge, this PGD approach represents the first attempt to construct a complete simulation of the global thermo-mechanical process. As is well known at this stage of the book, PGD uses a separated representation of the unknown field, in that case temperature or displacements, and results in a tremendous reduction of the computational complexity of the model. Moreover, it entails the ability to introduce any type of parameters (geometrical, material ...) as extra-coordinates of the model, to obtain by solving only once the resulting multidimensional model the whole envelope containing all possible solutions, i.e., a sort of numerical vademecum.

11.1 Parametric Modeling

The main idea that lies under the PGD is the separated representation of a field. Indeed, considering, for instance, an unknown quantity $u(\mathbf{x})$, it can be approximated by a finite sum of modes, as is done in modal analysis of structures, for instance. As previously discussed, the potential of the PGD goes far beyond the separation of space or time coordinates. Here, parameters considered as static input data in classical discretization techniques, can be introduced as extra-coordinates of the model. Let us imagine for instance that a field we would like to compute depends on a space coordinate $\mathbf{x}$, on time t, and on a set of Q parameters $\left(p^1, p^2, \ldots, p^Q\right)$. These parameters can be geometrical (thickness of a ply for example), material parameters (conductivity, reinforcement orientation, ...) Regardless of the nature of these parameters, the strategy is identical. The unknown field is sought under a separated form, also called finite sum decomposition,

$$u\left(\mathbf{x}, t, p^1, \ldots, p^Q\right) \approx \sum_{i}^{D} X_i(\mathbf{x}) \cdot T_i(t) \cdot P_i^1(p^1) \cdots P_i^Q(p^Q), \tag{11.1}$$

where $\mathbf{x} \in \Omega_x$, $t \in \Omega_t$ and $p^j \in \Omega_p$ for $j \in (1, \ldots, Q)$. Hence we obtain a muldimensional solution valid for any value of the parameters p^j within their domain of variability Ω_p supposed identical for the sake of simplicity. Even if the computation of such a solution can become heavy due to the high number of modes required, it can be done off-line once and for all. This "macro-solution" is then only particularized on-line, by specifying values for the parameters p^j. The associated computing time is very short which makes this approach quite promising for future real-time applications. Finally, the PGD allows us to build, for a given problem, a numerical abacus in which the model parameters act as extra-coordinates. In the next section dedicated to the thermal model for the ATP process, we explain how we can take advantage of this numerical tool to compute the temperature field in the composite laminate during forming.

11.2 ATP Thermal Model

As mentioned in the previous Introduction, the domain we consider is 3D and the material, the carbon reinforced PolyEther Ether Ketone (PEEK), is anisotropic. The thermal and mechanical properties of this material are detailed in Tables 11.1 and 11.2. In these tables, index 1 refers to the longitudinal or fiber direction, while index 2 corresponds to the transverse direction and index 3 stands for the "through-the-thickness" direction.

Table 11.1 Thermal properties of carbon reinforced PEEK (AS4/APC2)

Thermal diffusivity ($10^{-6}\,\mathrm{m}^2/\mathrm{s}$)	K_{11}	K_{22}	K_{33}
	1.89	0.189	0.189
Thermal expansion ($10^{-6}/\mathrm{K}$)	α_{11}	α_{22}	α_{33}
	0.2	60	60

Table 11.2 Mechanical properties of carbon reinforced PEEK (AS4/APC2)

Young Modulus (GPa)	E_{11}	E_{22}	E_{33}
	137	9.4	9.1
Poisson Ratio	ν_{12}	ν_{13}	ν_{23}
	0.33	0.32	0.40
Shear Modulus (GPa)	G_{12}	G_{13}	G_{23}
	5.1	4.7	3.2

Our objective is to obtain a steady state for the temperature in the system of coordinates related to the laser and the roller, that is to say the transient evolution assumed to be the same everywhere except in the vicinity of the edges. Indeed, far from the edges of the computational domain and for a given number of plies, each region of the material experiences the same thermal history during the process. It is progressively heated when approaching the laser, it reaches its maximum temperature at the location of the laser and it cools down relatively fast when getting far from the heat source. Therefore, instead of considering a problem where the domain is fixed and the boundary conditions are time dependent, we can explicitly introduce the line speed $\mathbf{v} = (v, 0)$ in the heat transfer equation by adding a convection term. In other words, the laser and the roller are kept fixed but the material is moved with a speed v (direction opposite to the advance direction of the laser in the process), as shown on Fig. 11.2. Hence the equation to be solved reads,

$$\rho C_p \left(\mathbf{v} \cdot \nabla T\right) = \nabla \cdot \left(\mathbf{K} \cdot \nabla T\right), \tag{11.2}$$

where ρ is the density, C_p the specific heat and $\mathbf{K}$ is the conductivity tensor. Employing this strategy, the boundary conditions are not time dependent anymore. The solution of Eq. (11.2) corresponds to the steady temperature field in the coordinate system attached to the consolidation roller and the laser.

The incoming material is assumed having ambient temperature. On the downstream boundary the heat flux is assumed vanishing. Convection boundary conditions are enforced on the upper surface, except in the region in which the laser applies, and finally a conduction transmission condition is enforced in the contact between the composite and the work plane. All the transmission conditions (inter-plies or composite-work-plane are affected by a resistance of contact h because of the non-perfect contacts. These resistances depend on the applied pressure and also on the inter-plies bonding. Thus, the temperature field becomes discontinuous at the plies interfaces and also on the composite-work-plane contact.

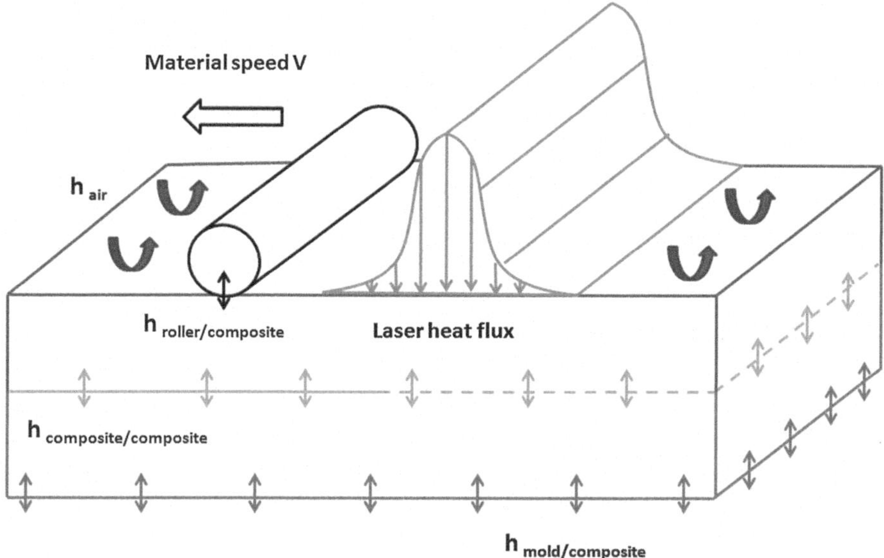

Fig. 11.2 Thermal model

11.2.1 Parametric Model

However, the thermal model presented until now is not fully satisfactory, because as soon as the sequencing of plies changes, the thermal model must be solved again because the thermal conductivities has changed.Thus, if we are interested in pre-computing all the possible sequencing, with 4 possible orientations and 10 plies, the number of possible sequences reaches the value of 4^{10}, even if the laser velocity of advancement, the laser power and the contact thermal resistances are considered static data.

One possibility consists of introducing the orientation of each ply as an extra-coordinate of the model [3]. By doing so, we obtain a solution valid for any ply sequence. The PGD separated representation allows circumventing the dimensionality increase. If we denote by θ^j the orientation of ply number j, the separated representation of the temperature field reads

$$T\left(x, y, z, \theta^1, \ldots, \theta^{N_p}\right) \approx \sum_{i=1}^{i=N} X_i(x, y) \cdot Z_i(z) \cdot \Theta_i^1\left(\theta^1\right) \cdots \Theta_i^{N_p}\left(\theta^{N_p}\right), \quad (11.3)$$

with $(x, y) \in \Omega_{xy}$, $z \in \Omega_z$ and $\theta^j \in \Omega_\theta$ for $j \in \left(1, \ldots, N_p\right)$ where N_p is the total number of plies in the laminate.

Several remarks can be made regarding approximation (11.3). First of all, for the sake of simplicity we assume that the orientation of each ply takes its values in the same domain Ω_θ.

The multidimensional solution $T\left(x, y, z, \theta^1, \ldots, \theta^{N_p}\right)$ is computed off-line, once for life. This numerical abacus will subsequently be used for the solution of the thermo-mechanical problem. It is important to highlight that velocity v, laser power and contact thermal resistance h can be included as extra-coordinates in order to obtain, always off-line, a more complete parametric solution [4].

11.3 ATP Mechanical Modeling

The solution of the thermal problem was a preliminary step in the progression toward our final goal: the evaluation of the process-induced residual stresses in the Automatic Tape Placement. The strategy we employ differs from the one we used for the thermal problem. Previously, we were working a coordinate system attached to a fixed laser/roller in the presence of moving material, as depicted in Fig. 11.2. As far the mechanical model is concerned, we consider the laser/roller moving with line speed v on the top surface of the laminate supposed now at rest.

One of the main difficulty is due to the change of geometry of the domain during the process. As the laser and the consolidation roller moves along the placement direction, the interface between the tape being placed on the substrate evolves: the welded region grows. The geometry evolution induces the installation of residual stresses in the laminate during the cooling process.

11.3.1 Interface Treatment

For the sake of clarity but without loss of generality, let us focus on a simple configuration: one ply being placed on a substrate composed of a unique ply. These two plies do not have necessarily either the same thickness nor the same orientation. In order to take into consideration the evolution of the interface between both plies during the process, we introduce a fictitious interface which extends along the placement direction. Numerically, this is done by duplicating the nodes along the interface that are attached as soon as the roller reach their location. We obtain therefore two distinct domains, one corresponding to the tape being placed Ω^+, and another one corresponding to the substrate Ω^-. The problem to be solved in each domain is identical and reads,

$$\begin{aligned} \nabla \cdot \sigma &= 0, \\ \sigma &= \mathbf{C}\left(\varepsilon - (T - T_0)\,\alpha\right), \\ \varepsilon &= \frac{1}{2}\left(\nabla \mathbf{u} + (\nabla \mathbf{u})^T\right), \end{aligned} \qquad (11.4)$$

where σ stands for the stress tensor, $\mathbf{C}$ is the fourth-order elastic tensor, ϵ is the linearized deformation tensor, $\mathbf{u}$ is the displacement field, T_0 is a reference temperature and α is the thermal expansion tensor. But, the boundary conditions to be used vary from one domain to the other. See [4] for additional details.

As for the thermal problem, the PGD is used. Space coordinates are separated in the same way: the in-plane coordinates, x and y, are kept together whereas z is treated separately. Instead of solving a 3D problem we only have to solve a few 1D and 2D problems. Actually, the computational complexity of 1D problems being negligible, the PGD allows us to solve a problem originally defined in 3 dimensions with the computational complexity characteristic of a 2D problem. See [5] for a deep analysis and discussion. Nevertheless, in the present analysis the fiber orientation of each ply of the laminate was not introduced as an extra-coordinate of the model, but it could be done exactly as we previously explained. Thus the displacement fields is written:

$$\begin{aligned} u(x, y, z) &\approx \sum_{i=1}^{i=N} X_i^u(x, y) \cdot Z_i^u(z), \\ v(x, y, z) &\approx \sum_{i=1}^{i=N} X_i^v(x, y) \cdot Z_i^v(z), \\ w(x, y, z) &\approx \sum_{i=1}^{i=N} X_i^w(x, y) \cdot Z_i^w(z), \end{aligned} \qquad (11.5)$$

where u, v and w are the displacements along the x-, y- and z-axis respectively.

11.3.2 An Incremental Approach

The strategy employed to evaluate the process-induced residual stresses uses the concept of representative volume. The objective is to determine a region of the domain, far enough from the left and right boundaries (placement direction) in order to avoid any edge effects and to determine a representative stress history at each position.

Starting from an initial configuration, which corresponds to the roller and the laser located at the left boundary of the representative volume, both are progressively moved to the right side in an incremental way (see Fig. 11.3): the model is thus quasi-static. Then, for a given position of the roller/laser, the mechanical problem (11.4) is solved and uses the temperature field previously computed. Once the displacement field for a given configuration has been computed, the problem is updated by modifying the position of the laser and thus the temperature field and the updated interface as illustrated in Fig. 11.3 that consider a damage variable to represent the interface.

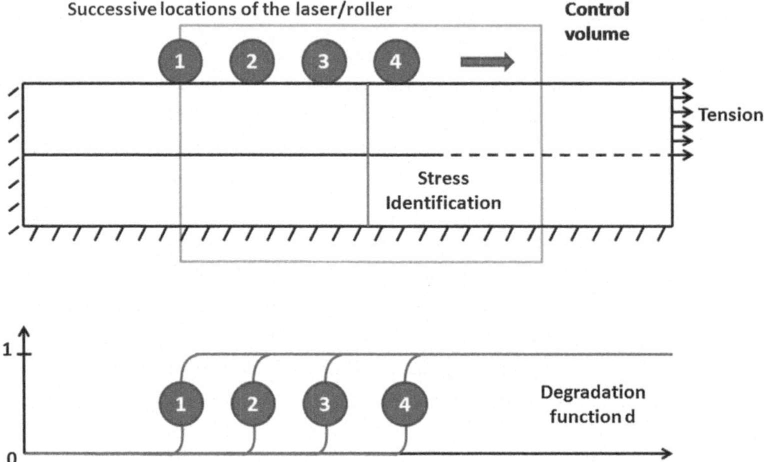

Fig. 11.3 Incremental approach

In order to evaluate a representative stress history in the laminate induced by the process, we extract at each increment the value of the stress at a given location, corresponding to the central cross-section of the representative volume (red line in Fig. 11.3). Doing so, it's then possible to capture the progressive increase of the stress when the laser gets closer resulting from the thermal expansion behavior of the two plies which tend to be incompatible, in particular if their fiber orientation is different. A maximum value is reached when the laser is at the exact location of "observation", and then it decreases as the device moves forward. For the last increments, the stress reaches a final value. This value corresponds to the residual stress in the laminate.

11.4 Numerical Results

Two very simple configurations are analyzed. On the one hand a unidirectional 2 plies laminate whose fiber orientation is parallel to the placement direction (x-axis in the global coordinate system). On the other hand a [90/0] laminate, commonly called cross-ply laminate, in which the carbon fibers in the incoming tape are oriented along the y-axis and are therefore orthogonal to the fibers of the substrate. Since the main source of stress in the laminate is the thermal expansion due to the laser heat flux, we could expect a priori a higher value for the residual stress in the second configuration. Indeed, because the orientation from one ply to the other changes, each ply deforms differently when experiencing a similar heating.

The thermal and mechanical characteristics considered in the numerical model are summarized in Tables 11.1 and 11.2. The dimensions of a ply are: length $L = 3\,\text{m}$, width $W = 0.2$ and thickness $H = 0.001\,\text{m}$

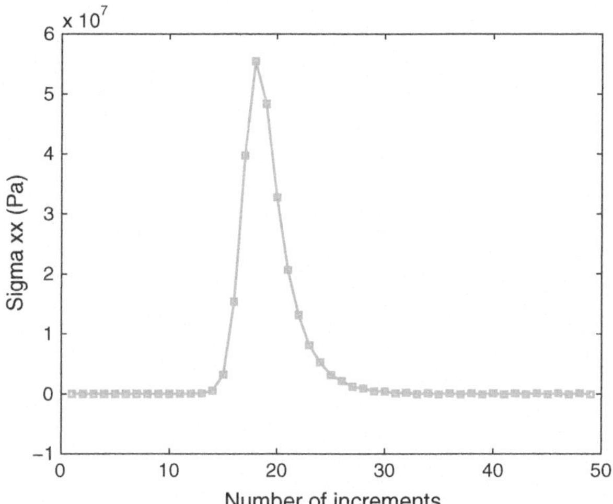

Fig. 11.4 Evolution of the stress σ_{xx} in the unidirectional laminate at $z = 5.10^{-4}$ m

11.4.1 Unidirectional Laminate

As expected, the evolution of the stress within the laminate consists of three different phases, as we can observe in Fig. 11.4. During the first phase, the laser is far away and therefore the laminate experiences very low stresses: because of the low thermal conductivity of the material, the thermal expansion is very localized. In the second phase, the laser is close enough to induce a thermal expansion: the stress increases and reaches a maximum when the position of the laser coincides with the observation point. Finally, the stress decreases gradually as the laser moves forward and tends to reach a constant value. This final value corresponds to the residual stress induced in the laminate by the thermal expansion and the welding of the two plies. Since both plies have the same orientation the resulting residual stress is quite low as illustrated in Fig. 11.5.

11.4.2 Cross-Ply Laminate

Since in this case the orientation is different in each ply, the residual stress in the laminate is expected to be much higher. The cross-ply laminate represents one of the worst configurations since the fibers in each ply are orthogonal to each other. The thermal expansion in the fiber direction is very low compared to the transverse direction. In our model, the ratio $\frac{\alpha_{22}}{\alpha_{11}} = 300$. Thus, although the temperature field in each ply is very similar, due to their small thickness, the resulting thermal expansion differs a lot. At the interface, the behavior of each ply is incompatible since the 0

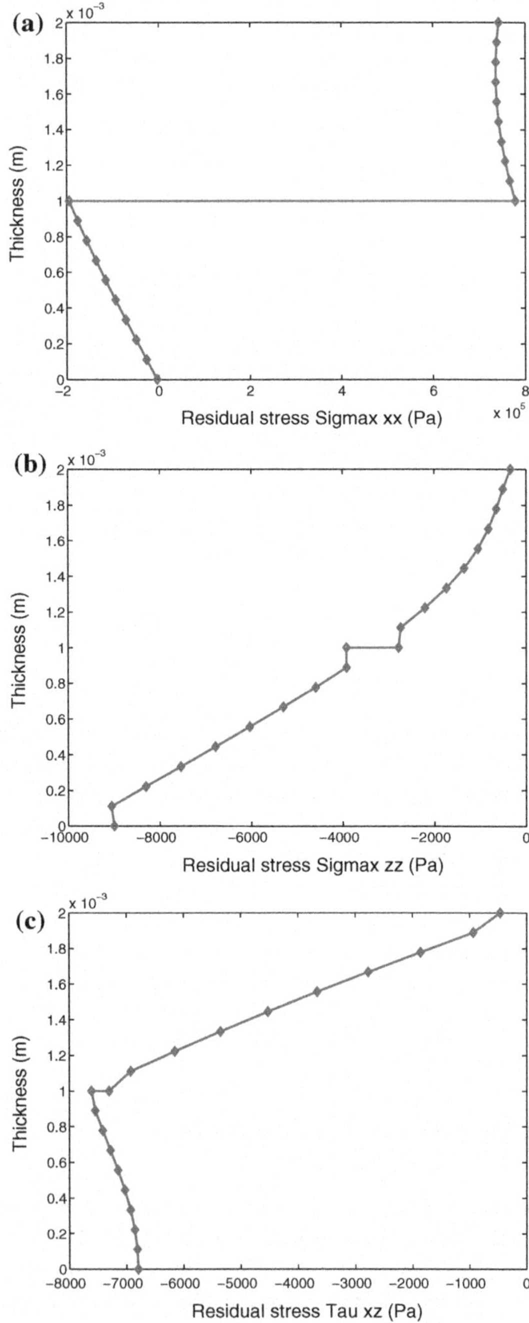

Fig. 11.5 Residual stress distribution through the laminate thickness **a** σ_{xx} **b** σ_{zz} **c** τ_{xz}

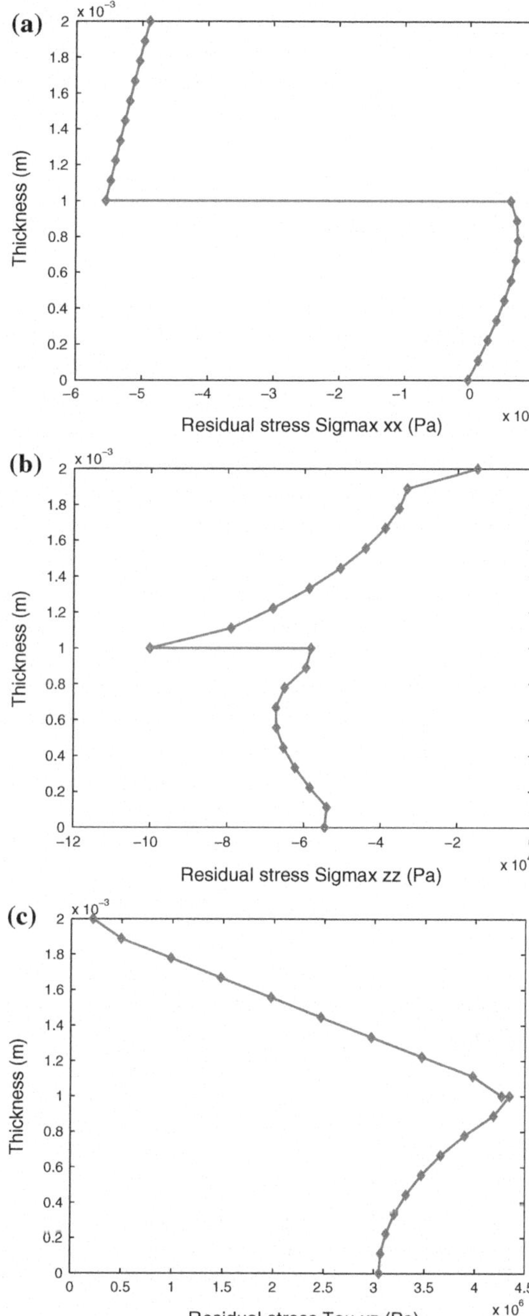

Fig. 11.6 Residual stresses distribution through the cross-ply laminate thickness **a** σ_{xx} **b** σ_{zz} **c** τ_{xz}

degrees ply expands along the y and z directions whereas the 90 degrees ply expands along the x and z directions. The welding leads to an important deformation at the interface. The residual stress in the laminate is then higher than for the unidirectional laminate as noticed in Fig. 11.6.

11.5 Conclusions

Evaluation of residual stresses induced by the automated tape placement process requires three distinct steps. The temperature field in the laminate has to be first calculated. Several approaches are conceivable. Here, we solve the thermal problem in the coordinate system attached to the heating device. The line speed is therefore explicitly introduced in the formulation of the problem by adding a convection term. The Proper Generalized Decomposition proceeds by decoupling the space coordinates by performing an in-plane/out-of-plane decomposition that allows solving the 3D problem with the computational complexity characteristic of 2D solutions. Moreover, the fiber orientation of each ply can be introduced as an extra-coordinate of the model and a parametric solution valid for a large range of laminates (sequencing of plies) can then be computed. Other extra-coordinates can be also introduced allowing efficient material and process identification and/or optimization.

The mechanical problem is solved incrementally in a representative volume. The laser moves progressively along the placement direction. For a given position, a thermo-mechanical problem is solved, making use of the temperature field already computed. The residual stress is obtained by considering the evolution of the stress on the central cross-section of the representative volume. Finally, with the residual stresses just obtained, the part can be demolded and the induced distortion can be calculated by solving the associated elastic problem at the structure level again by invoking the in-plane-out-of-plane PGD decomposition of the associated elastic problem.

Nevertheless, very simple configurations were analyzed and this work has still to be validated with more complex simulations and experiments that constitute a work in progress.

References

1. F.O. Sonmez, H.T. Hahn, M. Akbulut, Analysis of process-induced residual stresses in tape placement. J. Thermoplast. Compos. Mater. **15**, 525–544 (2002)
2. R. Pitchumani, S. Ranganathan, R.C. Don, J.W. Gillespie, Analysis of transport phenomena governing interfacial bonding and void dynamics during thermoplastic tow-placement. Int. J. Heat Mass Transfer **39**, 1883–1897 (1996)
3. C. Ghnatios, F. Chinesta, A. Barasinski, A. Leygue, A. Poitou. Modelisation et simulation 3D parametrique des champs de temperature dans des plaques composites, Conference MECAMAT, Aussois (2011)

4. F. Chinesta, A. Leygue, B. Bognet, Ch. Ghnatios, F. Poulhaon, F. Bordeu, A. Barasinski, A. Poitou, S. Chatel, S. Maison-Le-Poec. First steps towards an advanced simulation of composites manufacturing by Automated Tape Placement. Int. J. Mater. Form. (2006). DOI:10.1007/s12289-012-1112-9
5. B. Bognet, A. Leygue, F. Chinesta, A. Poitou, F. Bordeu, Advanced simulation of models defined in plate geometries: 3D solutions with 2D computational complexity. Comput. Methods Appl. Mech. Eng. **201**, 1–12 (2012)

Chapter 12
Augmented Learning via Real-Time Simulation

If you can't explain it simply, you don't understand it well enough.

—Albert Einstein

This chapter deals with an innovative approach to engineering learning. It will be shown how engineering education could eventually benefit from the ability of PGD to construct efficient algorithms for the real-time simulation of complex physics. This fact, and its ability to be implemented in handheld, deployed platforms, to rend an interactive tool, opens the possibility to develop very efficient *augmented* learning strategies for higher education.

In higher education contexts, and noteworthy in engineering and applied sciences, it has been found that simulation-based learning enhances not only self-determined motivation, but also improves learning in general. Furthermore, *augmented learning* has emerged as a new way to improve the learning process by adapting the environment to the learner. This is achieved, notably, with the use, among other devices, of touchscreens or haptic peripherals. Some authors claim that these results can be obtained even in a non-conscious manner.

Simulation allows teachers to reproduce complex physics otherwise impossible to reproduce experimentally in the classroom. Going one step ahead, handheld devices add their portability, interactivity and sociability to the simulation framework. Also the editorial world is suffering drastic changes with the irruption of enhanced media as classroom material.

In the field of higher education, however, the development of augmented learning tools becomes still more challenging. Some apparently very simple equations (such as the Schroedinger equation, for instance) can not be simulated with nowadays computer capabilities.

The task of implementing augmented learning strategies for complex engineering or physical problems is therefore a formidable challenge. These problems take days of supercomputing facilities to be solved. The simple possibility of solving them on deployed, handheld platforms seems to be out of reach. In addition, traditionally,

F. Chinesta and E. Cueto, *PGD-Based Modeling of Materials, Structures and Processes*, ESAFORM Bookseries on Material Forming, DOI: 10.1007/978-3-319-06182-5_12,

simulation-based engineering sciences make use of static data. By *static* we mean here that inputs can not be changed dynamically during the course of the simulation. But in augmented learning environments we face a clear example of Dynamic Data Driven Application Systems (DDDAS). Nowadays, the linkage of simulation tools with external dynamic data for real-time control of simulations and applications is becoming more and more frequent. In the educational context, the external dynamic data is given by the learner's interaction instead of measurement devices. DDDAS constitute one of the most challenging applications of simulation-based engineering sciences, due to the dynamic incorporation of additional data into an executing application, and *vice versa*, the ability of an application to dynamically steer the interaction process [1].

One of the most challenging applications in the field of DDDAS (also in the field of augmented learning) is that of *haptic* systems. By haptic we refer to those peripherals in an informatics system that provide the user with tactile sensations. While in cinemas, to obtain a perception of continuous movement some 25 photograms per second are needed, haptic devices need for 500 Hz-1 kHz feedback rate to give a physically accurate sensation of touch. That is, haptic devices need a feedback rate that is 20 to 40 times faster than the cinema one, and instead of sending a predefined image may require some computations.

In this chapter we analyze how PGD can help in the development of augmented learning strategies for applied sciences and engineering higher education.

The main interest from the perspective pursued in this work is that PGD methods opened the door to an innovative treatment of classical problems, not necessarily in high dimensions. When dealing with parametric problems, for instance, it has been seen that these parameters can advantageously be considered as new coordinates of the problem. Since the PGD is able to cope with an arbitrary number of dimensions without the typical exponential increase in terms of degrees of freedom, this procedure is now at hand.

By considering parameters as new spatial dimensions, a new class of solutions is obtained for this type of problems. The PGD method is therefore able to obtain solutions for *any* value of the parameters at *any* point of the space, at *any* time instant. So to speak, PGD provides the analyst with a sort of meta-model, but with the strong novelty of no need for previous computer experiments. Many problems in science and engineering can thus be analyzed once and for all, since the method provides the solution for any value (in a given interval) of the parameters.

This simple approach has been implemented in a two-stage approach. The first stage is concerned with the *off-line* computations in which the general form of the solution is computed, and as mentioned before once and for all. This phase may take intensive computing resources, and may need for supercomputing resources, parallel computing, GPUs, ... Note, however, that in general the computer cost at this stage is not critical since it will be done only once. Then, with the results obtained in the first off-line stage, an *on-line* phase can be accomplished. At this on-line stage the user, namely the learner in an educational context, can make use of the results in real-time, even on deployed, handheld devices, thanks to the simplified structure of the approximation.

12.1 Towards Simulation-Based and Augmented Learning for Complex Physics: iPGD

Based on the two-stage strategy presented before, one off-line intensive computing phase followed by a fast real-time on-line simulation phase, the authors have developed and implemented a PGD platform able to run on smart-phones and tablets running Android 3.0 or higher operative systems. It is called: iPGD.

The application, that can be downloaded freely, see [2], makes use of a slight modification of the *eXtensible Data Model and Format* (xdmf) for the input files. In it, the different functions involved in the separated representation are usually stored as one-dimensional arrays of values, thus minimizing the needed memory usage. Note that the Android 3.0 operative system precludes the use of more than 65536 points (nodes) in the model. This is avoided by using meshes of the boundary of three-dimensional solids only, instead of a fully 3D-mesh. The iPGD application is able to cope with real and complex-valued fields. In Fig. 12.1 a screenshot is shown on the appearance of the application. Moreover, lighter versions have been implemented on Nokia phones with similar outputs, see Fig. 12.2. When running on a handheld device such as an smartphone or tablet, the iPGD application is able to provide the user with feedback rates enough for visual realism (on the order of 30 Hz). If haptic feedback rates are needed (i.e., on the order of 500 Hz), the application must be run on a laptop (tested on a MacBook pro running OS X Lion).

12.2 Examples of the Performance of the Proposed Technique

In this section several examples of how the iPGD application is being used advantageously are given. In all cases we deal with complex physics, difficult to understand for undergraduate students of several disciplines. The iPGD application provides them with different skills ranging from a physical interpretation of complex-valued equations, as is the case for the Helmholtz equation governing the wave agitation in a harbor, to anatomy and (possibly) soft tissue stiffness for biomedical applications in the field of endoscopic surgery.

12.2.1 Plate and Shell Structures for Engineers

Plates and shells are structural elements with one dimension considerably smaller than the other two. They are subjected to bending and stretching (membrane) efforts. The hypothesis usually employed in structural engineering to model them induce several difficulties from the theoretical as well as from the practical and learning points of view. Two major theories exist. On one hand, there is the Kirchhoff-Love theory which neglects the shear influence on the displacement (rotation) of the cross-sections

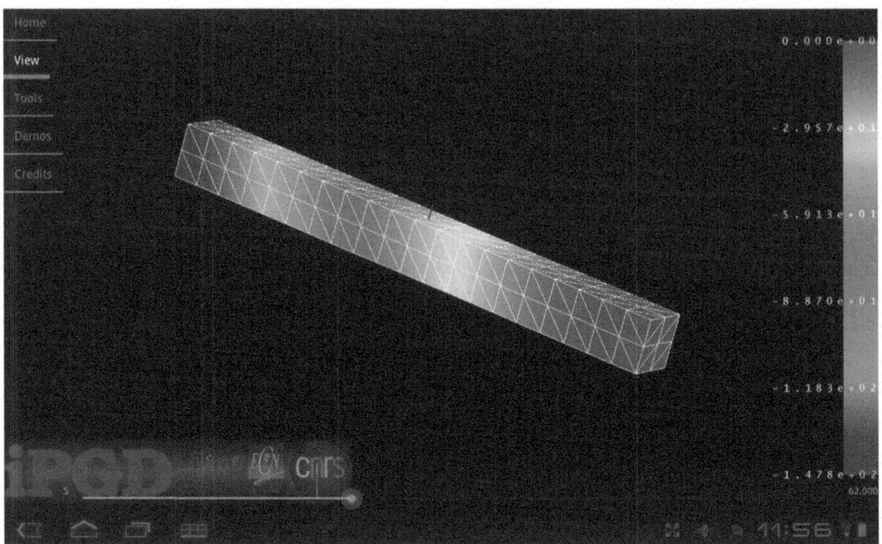

Fig. 12.1 Appearance of the iPGD Android application

Fig. 12.2 The iPGD application running on a Nokia smartphone. In this case an industrial furnace thermal simulation is running. The learner can manually adjust the working temperature on the heaters to observe the effect on the resulting temperature field

and therefore assumes a flat section after the deformation. This theory is assumed valid for thin plates. On the other hand, there is the Reissner-Mindlin theory which assumes an additional rotation of the cross sections due to the shear stresses.

Plate and shell theory imply concepts that are non-trivial and induce difficulties for undergraduate students in engineering. In fact, acquiring the required skills in this field is a major challenge in engineering education. The relationship of the kinematic assumptions to the physical behavior of actually three-dimensional solids is not straightforward to understand and, in addition, these theories have well-known limitations and inconsistencies. For instance, Kirchhoff-Love theories have been corrected to give appropriate values of the reactions for simply supported or clamped plates, for instance.

As a matter of fact, the PGD approach presented before constitutes an unprecedented method for the interactive learning of plate and shell theories. The PGD approach to the problem [3] does not assume any kinematic or static hypothesis on the behavior of plates and shells. In fact, it simply performs a full 3D calculation (a 2D plus a 1D calculation indeed, in separated form) in which the thickness of the plate, h, is not a parameter but an additional coordinate of the model:

$$u(x, y, z, h) \approx \sum_{i=1}^{i=N} F_i^1(x, y) \cdot F_i^2(z) \cdot F_i^3(h). \tag{12.1}$$

Thus, the student is able to dynamically change the thickness of the plate without changing the model to see the actual effect on the overall plate response. This is achieved by just particularizing the solution given by Eq. (12.1) to the selected value of plate thickness, h.

The separated structure of the solution provides a very interesting learning capability for this special case, since the student can notice how the through-the-thickness solution of the first functional pair of the solution varies from a straight line to more complex structures, depending on the chosen thickness of the plate, see Fig. 12.3. In Fig. 12.4 one can notice how a correction to Kirchhoff-Love theories near the clamped boundaries is necessary. This is achieved by just considering the thickness of the plate as an additional parameter (dimension) of the solution at the off-line phase of the PGD.

All these details can be consulted dynamically by the student on his smartphone, for instance. In Fig. 12.5 an example of the effect of adding several plies of different (possibly orthotropic) materials to the plate is studied. The Android version of the implementation is somewhat more sophisticated, see Fig. 12.6.

12.2.2 Wave Height on a Harbor

Knowledge of sea-wave propagation is crucial for many stages of harbor design. For instance, an optimal shape is required to minimize the wave agitation in some zones of interest, usually those areas where more shipping activities are concentrated. Understanding how the incident wave can amplify inside the harbor is a major priority, for example when resonance conditions are of concern. Furthermore, the wave behavior while propagating under complex harbor models is not an easy task to be interpreted by engineers, see Fig. 12.7. The standard model which governs the wave

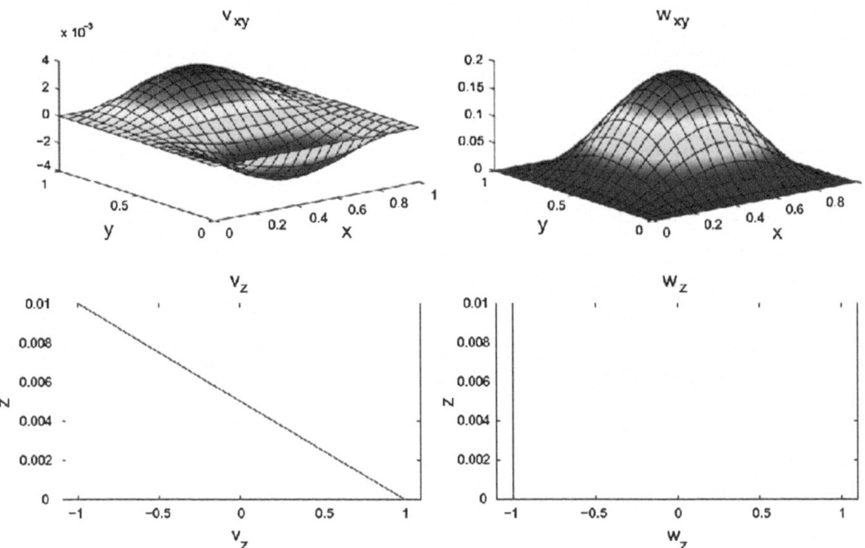

Fig. 12.3 Solution provided by the PGD method. The first pair of functions F shows the relatively good accordance of the solution with respect to Kirchhoff-Love assumptions

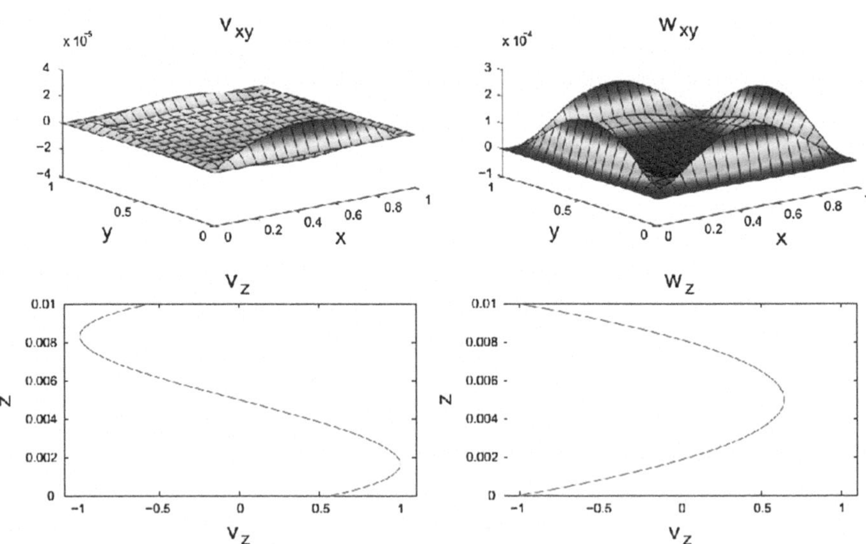

Fig. 12.4 The second pair of functions shows clearly how Kirchhoff-Love theory needs for a corrections near the clamped boundary

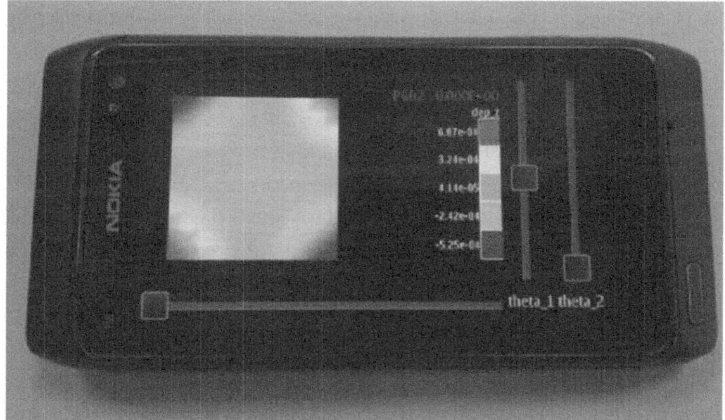

Fig. 12.5 An example of the implementation of plate and shell theories on a smartphone. The learner can adjust dynamically the orientation of different plies on the overall behavior of the plate

Fig. 12.6 An example of the implementation of plate and shell theories on the Android platform

Fig. 12.7 Example of simulation of the wave propagation in the Barcelona harbor

propagation in harbors is the two-dimensional Helmholtz equation in unbounded domains. Its solution provides the wave height, or wave amplification, when dealing with a single incident wave in terms of period (or wavelength) and incident direction of propagation.

But, the physical wave response on harbors extremely depends on these two model parameters. A complete wave agitation analysis requires the solution of the previous problem for any possible combination of incident wave periods and directions, determined by experimental measurements in the coastal area. Typical applications are control, optimization or inverse problems, where usually a large number of solutions are required, and, consequently, it is necessary to solve many direct problems. The solution of real harbor models demands important computational resources and usually implies extremely large computing times. Thus, usual optimization procedures are inapplicable, either because they need numerous solutions or because real-time constraints are required. To take charge of this issue, standard engineering procedure lies in solving the model with a certain set of parameters trying to cover the maximum number of combinations. Clearly, this "brute force" strategy imposes a bound in the final number of cases to be analyzed and, therefore, important loss of information becomes evident.

The presented PGD approach for this kind of problem involves, in separated form, the incident wave period and the incident direction as new coordinates in the model. With this strategy, the full 4D problem is reduced to the iterative solution of one 2D problem and two 1D problems. Hence, the solution of this model breaks the barrier that has prevented the full incident data evaluation up to nowadays. By means of readily evaluating, offline, any incident wave, engineers can obtain the wave propagation inside a harbor in a real-time manner, giving them a formidable support to better analyze where the wave amplifies for all the possible real situations.

At the same time, the solution given by the PGD approach in the Mataró harbor is implemented on the iPGD application for Android platforms, see Fig. 12.8. In this case, students can dynamically change the incident period and obtain the wave amplification, in real-time, inside the area of interest, with no need of high computational resources.

12.2.3 The Case of Anatomy and Endoscopic Training

Dissection is recognized as the best practice to learn gross anatomy. However, the price of this class of experimental training in medical universities becomes prohibitive in most cases, due to problems arising from logistic and financial issues. Teaching anatomy with the help of multimedia and computers is nowadays an extended practice, see [4], for instance. Some recent specialities, such as endoscopic (laparoscopic) and minimally invasive surgery, for instance, need even more sophisticated training.

In endoscopic training, surgeons have been trained for years with the help of animal models, i.e., by experimenting on pigs, for instance. Apart from being very expensive, this type of training does not fit with the modern ethical treatises about

Fig. 12.8 Example of the use of the iPGD for the wave propagation in the Mataró harbor with any incident period between 5 and 14 seconds

experimentation on animals. Instead, simulations provide learners with an immersive environment in which a versatile change of scenario is possible where even extremely rare cases can be studied.

The development of surgery simulators with haptic feedback is an active area of research due to the complexity of the problem. One of the sources of complexity is due to the highly non-linear behavior of soft living tissues, that are frequently modeled under the fiber-reinforced hyperelasticity framework. Other source of complexity comes from the highly restrictive feedback rates imposed by the simulators (25 Hz for visual feedback and some 500 Hz if we want to add haptic feedback to the system). The third source of complexity comes from the multi-physic nature of the phenomena occurring in the actual surgery procedure: non-linear elasticity, contact, cutting, temperature, etc.

Such simulators should provide a physically more or less accurate response such that, with the use of haptic devices, a realistic feedback is transmitted to the surgeon in terms of both visual feedback and force feedback. By "accurate response" we mean that an advanced user should not encounter "unphysical" sensations when handling the simulator. We definitely do not pursue an accurate solution in engineering terms. Following [5], "... the model may be physically correct if it looks right".

Ayache et al. [6] defined the three generations of surgical simulators as those able to, respectively, reproduce accurately the anatomy (geometry), the physics (constitutive equations of soft tissue, temperature) and, finally, physiology (blood flow, breathing, ...). Undoubtedly, no third-generation simulation has ever been developed, and only some rigorous attempts have been made at the second-generation level. Most of the existing simulators can be classified into the first-generation category, even if they provide haptic feedback, because they only consider linear elastic response. Other are based upon spring and mass systems, that do not even reproduce the equations of linear elasticity (see [7] and references therein). These approaches are judged clearly insufficient and clearly non-realistic by most surgeons [7].

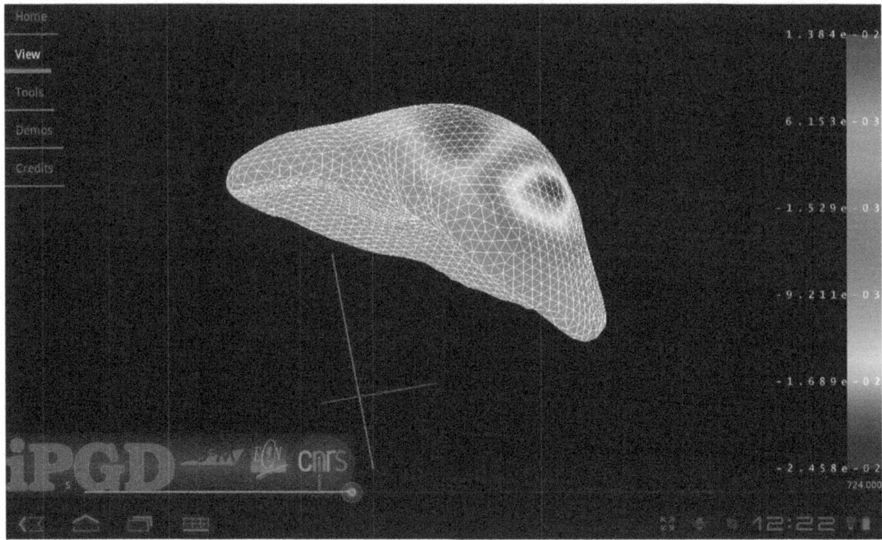

Fig. 12.9 An example of the use of the iPGD application in anatomy. In this case a human liver is being studied

The proposed method can be used at two different levels. Firstly, if one considers a medical student, freshman or sophomore, the iPGD application introduced before offers the possibility of an augmented learning strategy in which not only the geometry (anatomy) of the human body can efficiently be displayed, but also the learner can dynamically interact with the tissues and their relative stiffness by tapping on the screen of the handheld device (even if no haptic experience can be obtained at this level). In this way an improved experience over traditional methods [4] can be obtained, see Fig. 12.9.

On the other hand, if connected to appropriate haptic devices, a standard computer and monitors, an operating theatre can be simulated efficiently. No special computational resources are necessary in this case for the on-line phase of the simulation, and even complex soft tissue models can be efficiently reproduced with haptic feedback and possibly augmented reality scenarios. In any case, the on-line phase of the simulation can incorporate complex, state-of-the-art constitutive laws without affecting the performance of the simulation [8–11], since the most part of the computer cost is paid at the off-line phase. The simulator is thus fed with the results of these simulations, possibly very time consuming, but made only once for life.

12.3 Conclusions

In this chapter we have introduced and studied the possibilities offered by proper generalized decomposition methods in the field of simulation-based and augmented learning. The PGD method constitutes a new paradigm in simulation-based engineering sciences, but also possesses salient features in the field of augmented learning. Notably, we have explored its ability to provide for effective solutions (in terms of computing time and also storage needs) for complex physics and engineering problems that have traditionally fall out of reach for traditional simulation techniques at real time feedback rates.

Several examples have been provided that show the potential of the proposed techniques to offer real-time simulation scenarios enabling augmented learning to go beyond its present limits. These examples range from problems of the theory of solid mechanics to anatomy and virtual surgery training of surgeons.

Although it is envisaged that this technique could have a strong impact on the type of problem for which augmented learning strategies are developed, it is still necessary to further study how the approaches here presented could effectively improve the learning process of students at higher education institutions. Since the field of application is very broad, this type of study will probably need a detailed campaign of surveys, different for each speciality. For the time being, this approach is being employed in three different master-level courses at Ecole Centrale Nantes, France (twenty students in the computational materials course, 10 more at the modeling of composite manufacturing processes course and 12 students at the Erasmus Mundus Master program run jointly by EC Nantes, Swansea University, U.K., Stuttgart, Germany, and UPC BarcelonaTech, Spain).

References

1. NSF Final Report, *DDDAS Workshop 2006* (Arlington, VA, USA, 2006)
2. C. Quesada, D. Gonzalez, I. Alfaro, F. Bordeu, A. Leygue, E. Cueto, A. Huerta, F. Chinesta. Real-time simulation on handheld devices for augmented learning in science and engineering. PLOS ONE (Submitted)
3. B. Bognet, A. Leygue, F. Chinesta, A. Poitou, F. Bordeu, Advanced simulation of models defined in plate geometries: 3D solutions with 2D computational complexity. Comput. Methods Appl. Mech. Eng. **201**, 1–12 (2012)
4. M.J. Inwood, J. Ahmad, Development of instructional, interactive, multimedia anatomy dissection software: a student-led initiative. Clin. Anat. **18**(8), 613–617 (2005)
5. H. Courtecuisse, H. Jung, J. Allard, C. Duriez, D.Y. Lee, S. Cotin. Gpu-based real-time soft tissue deformation with cutting and haptic feedback. Progress in Biophysics and Molecular Biology, **103**(2-3),159–168 (2010)
6. H. Delingette, N. Ayache, Hepatic surgery simulation. Commun. ACM **48**(2), 31–36 (2005)
7. H. Delingette, N. Ayache, Soft tissue modeling for surgery simulation, in *Computational Models for the Human Body*, ed. by N. Ayache, *Handbook of Numerical Analysis*, ed . by Ph. Ciarlet (Elsevier, Amsterdam, 2004), pp. 453–550

8. S. Niroomandi, I. Alfaro, E. Cueto, F. Chinesta, Real-time deformable models of non-linear tissues by model reduction techniques. Comput. Methods Programs Biomed. **91**, 223–231 (2008)
9. S. Niroomandi, I. Alfaro, E. Cueto, F. Chinesta, Model order reduction for hyperelastic materials. Int. J. Numer. Meth. Eng. **81**(9), 1180–1206 (2010)
10. S. Niroomandi, I. Alfaro, E. Cueto, F. Chinesta, Accounting for large deformations in real-time simulations of soft tissues based on reduced order models. Comput. Methods Programs Biomed. **105**, 1–12 (2012)
11. S. Niroomandi, I. Alfaro, D. Gonzalez, E. Cueto, F. Chinesta, Real time simulation of surgery by reduced order modelling and X-FEM techniques. Int. J. Numer. Meth. Biomed. Eng. **28**(5), 574–588 (2012)

Index

F. Chinesta and E. Cueto, *PGD-Based Modeling of Materials, Structures and Processes*, ESAFORM Bookseries on Material Forming, DOI: 10.1007/978-3-319-06182-5,

The manufacturer's authorised representative in the EU is Springer Nature Customer Service Centre GmbH, Europaplatz 3, 69115 Heidelberg, Germany. If you have any concerns regarding our products, please contact ProductSafety@springernature.com

Printed and bound by CPI Group (UK) Ltd, Croydon, CR0 4YY
15/07/2026
02167631-0003